W9-ATH-773
Wenham Mass. 01984
DISCARDED
JENKS LRC
GORDON COLLEGE

THE PREDATORY BEHAVIOR OF WILD CHIMPANZEES

THE PRIMATES

THE PREDATORY BEHAVIOR OF WILD CHIMPANZEES

Geza Teleki

Lewisburg
BUCKNELL UNIVERSITY PRESS

Associated University Presses, Inc.
Cranbury, New Jersey 08512

Library of Congress Cataloging in Publication Data
Teleki, Geza, 1943–

The predatory behavior of wild chimpanzees.

Bibliography: p.

1. Chimpanzees—Behavior. 2. Predation (Biology)
I. Title.

QL785.5.C5T44 599′.884 70-124442
ISBN 0-8387-7747-3

Printed in the United States of America

To Hugh

and all those primates at Gombe
who accepted me as I was and came to be

CONTENTS

FOREWORD

The Gombe National Park, Tanzania, site of Geza Teleki's study of chimpanzee predation, is well known throughout the scientific world. The development of a research center in the Gombe Park dedicated to studying the naturalistic behavior of chimpanzees is the primary responsibility of Dr. Jane van Lawick-Goodall. Her achievements of creating field conditions and methodologies equal in importance the splendid contributions that she has made to the understanding of the biology, ecology, and behavior of chimpanzees. Dr. van Lawick-Goodall and her colleagues have made an enduring contribution to modern anthropological theory and perspective of mammalian primates and human evolution. She was the first to observe and to report objectively on chimpanzee predation, and it is within the context of the Gombe ecosystem and the center's historic development that Geza Teleki found very favorable conditions for this thorough and pace-setting report of the predatory behavior of Gombe chimpanzees.

There are two other important conditions affecting this study: With unusual liberality the Anthropology Department and the Graduate School of The Pennsylvania State University arranged academic accreditation and remote supervision for Teleki's field work. The second condition is the library of first-order notes, reports, research data and films that has been collected systematically, organized, and maintained on the Gombe study community of chimpanzees. This data base has made available the harvest of ten years of valid observations. The assistance of capable, cooperative observers has resulted in a far more complete and systematic account of chimpanzee predation than would have been possible with a single observer working however diligently for twelve months. The Gombe pattern of systematic and sustained observations is similar in some respects to the study programs of the Japan Monkey Center at Inuyama and to those of the Santiago

Island Colony, of the Caribbean Primate Research Center, Puerto Rico.

I believe that this book on anthropoid predation and others to follow on a great range of subjects of equal scientific merit will demonstrate the efficacy of developing field study conditions which include early risks, uncertainties, gradual and solid growth, associated with simplicity of living and working conditions, limited institutional support and remote control, and strong themes of research involving cooperation and mutual commitments of able people to defined objectives and their achievements. Another feature is the detached respect shown by observers for the primates being studied and the protection of the ecosystem of the observers and the observed.

Dr. Jane van Lawick-Goodall, Geza Teleki, the Gombe Stream National Park and chimpanzee predation are clusters of concepts which will long persist in the literature of the biological and behavioral sciences. The evidence is adequate for acceptance of predation in the chimpanzee behavioral repertoire, contingencies of its occurrence, conditions of its expression, and variations of all phases of the predatory action system. Years of additional observations and intensive study may be required to chart the full range of permutations and variations of primate predation and to learn more of chimpanzee predation in relation to its sociobiotic and evolutionary significance. Certainly chimpanzees in the Gombe study community are capturers and killers of the young of other mammals. The chimpanzee has been described as an omnivore. The kills and divisions of prey are dramatic, and Teleki is to be commended for cool restraint and utter objectivity in describing this behavior which creates images in the brains of the thoughtful of the early dawn of humanoid evolution, but what follows may be even more significant for sociocultural evolution. The eating of a prey object is a system of integral activities which are orderly and reflective of the social characteristics of the interactive chimpanzee clusters and subgroupings. The behavior mirrors their respective status, qualities of ownership, sharing, cooperation, and "mutual aid." Furthermore, the images of ritualistic and ceremonial behavior urge to be included in the descriptions, analyses, and interpretations of this phase of predation episodes. Those scientists and scholars who search for the *anlagen* or germinal origins of sociocultural transmissible behavior are invited to consider in all aspects the terminal acts of chimpanzee predation and their significance as regulators and indicators of community structure.

Even with the information collected during ten years, the story of chimpanzee predation is only beginning to unfold. The growth and spread by learning of predatory activities in the Gombe chimpanzees have not yet run their course! Both adult males and females and young may be learning to catch, kill, divide, and eat young animals of other species. A grave question intrudes: When will chimpanzees kill young of their own species, and under what conditions, and thus express another humanoid potential? Will the victim be a member of the same or another community? There are other questions: What are the physiological, nutritional, appetitive, and motivational factors which regulate the occurrence of chimpanzee predation? What are the effects of "provisionization," and the consequent crowding of chimpanzees and baboons, on predation? What are the levels of effectiveness of sign and signal inhibitors of attacks and aggression on young within species and across species? What are other similar kinds of behavioral regulators of predation? Does Teleki's vacant niche hypothesis have validity? Can predation be experimentally produced in captive chimpanzees and the results analyzed? What are the adaptive and nonadaptive functions of predation, and how are the effects selectively operative on primate populations and evolutionary changes? What is the role of predation in the evolution and regression of primates, including man? These and other questions invite extended investigations of primate predation along with other systematic studies of chimpanzees.

Although predation has been reported recently for chimpanzees in other communities, there is need for sufficient comparability of observational procedures and other conditions to make possible comparisons among different chimpanzee communities and to learn how the conditions of the contingencies of predation differ among communities.

It is probable that since anthropoid predation has been so dramatic, well observed, and reported for the Gombe study community of chimpanzees, other field investigators will now focus on and record this kind of behavior for many other primates. Perhaps the capture, killing, and eating of prey are a more prevalent feeding activity than previous concepts of primate food preference permitted us to see and to acknowledge. This revised emphasis is a part of recent developments for studying and recording the complete ecosystem of undisturbed populations of primates living in their native habitats.

The story that Geza Teleki tells has characteristics that make it a pace-setting example of natural science literature. In an era when there is insecurity about the relative importance and interdependence of quantitative and qualitative methods and the analysis and reporting of data, Teleki describes broad patterns of behavior involving the simultaneous interactions of many chimpanzees following linked sequential action systems which run a course from drive to goal. These descriptions resemble realistic human drama performed on a broad stage, with each ape being a character that takes its role and acts its part. Also, at a time when ethological analyses would tend to reduce behavior to small elemental components, Teleki has chosen to present holistic episodes of activities. Thus the integrated, organized patterns of actions are depicted, as are their functional biotic significances. Gombe's observing, recording, and reporting system, as used by Teleki in writing the exemplary episodes, achieves a new and high quality level of naturalistic literature.

It is well at this point to recall that the definition of descriptive categories of observations, evidence, and data is a necessary precondition to the application of statistics and mathematics in data analysis.

When one surveys the growing body of information of nonhuman primates and considers the kinds and amounts of information required for making reliable and valid comparisons with man, his long past, and extended present living conditions, the Gombe efforts come to have another important parameter or dimension. The decade of research and the resulting accumulation of large amounts of informative data on the chimpanzee begin to be adequate for making important comparisons among nonhuman and human primates, and these comparisons emphasize the necessity for having large blocks of data on the species before valid comparisons can be made among them. The conclusion is that sustained and systematic observations on permanently protected and managed subpopulations and colonies of nonhuman primates like Santiago Island, Puerto Rico; Takasakiyama, Kyushu, Japan; and Mount Visoke, Rwanda, are essential conditions for significantly advancing the primatological branch of the biological and behavioral sciences.

C. R. Carpenter
Athens, Georgia
October 1971

ACKNOWLEDGMENTS

Financial support of my work with the Gombe Stream Research Centre, in the Gombe National Park of Tanzania, was provided by the National Geographic Society, and I express my gratitude to all the members of this organization. The information and photographs presented here are used with the permission of the National Geographic Society and the directors of the Gombe Stream Research Centre. The Pennsylvania State University provided a study grant and special graduate program which enabled me to conduct field work abroad. I wish to also thank the Government and National Parks for the opportunity to live and work in Tanzania.

For my participation in the Gombe field program, the deepest and most sincere acknowledgments are due L. S. B. Leakey, J. van Lawick-Goodall, and C. R. Carpenter. Dr. Leakey was instrumental in recommending and arranging my joining the field research staff as an assistant. With Dr. van Lawick-Goodall are all my thanks for her patience, encouragement, and guidance during my work with the research team, and especially for allowing access to the unique data comprising this volume. To Professor Carpenter, my advisor at The Pennsylvania State University, is due gratitude for the many letters and problems shared while on distant continents, and for the practical advice and moral support so essential to this work and author. I am also indebted to E. E. Hunt and R. A. Hinde for their comments and advice on improving the text of this report. Final thanks must go to Baron Hugo van Lawick for his assistance and counsel in many field dilemmas.

Further acknowledgment for the labor and material presented in the ensuing pages goes to Miss A. Sorem, Miss C. Gale, Miss D. Staren, Miss C. Clark, Mr. P. McGinnis, and other field associates. Mr. and Mrs. T. W. Ransom, whose extensive knowledge of two local baboon troops was made freely available from their notes and in discussions, contributed greatly to my understanding of an important prey species. Emotions of a more personal and enduring nature rest with the late Miss R. A. Davis, who lost her life in an accident while following chimpanzees in 1969.

Printed words of gratitude and recognition are, however, a mute understatement of debts and emotions. Many hands, including those of family and friends, contributed to this preliminary account of predatory behavior in wild chimpanzees.

INTRODUCTION

This report describes the predatory behavior documented during ten years of field research with a natural population of chimpanzees at the Gombe Stream Research Centre, in what has recently become the Gombe National Park of Tanzania. Emphasis is given the thirty predatory episodes observed in detail during a one-year study period from March 1968 to March 1969. Comparable data on other living primate species, with the partial exception of some field reports on olive and yellow baboons in East Africa, are still scarce. There is every reason, therefore, to treat this new information on chimpanzee behavior with caution, especially with regard to implications and comparisons, pending at least some later determination as to whether such behavior is or is not idiosyncratic in time and location.

The report consists of three major sections followed by a summary of all the data gathered and a brief discussion of some problems relevant to the evolution of predatory behavior among primates. Chapter 1 outlines the environment of Kigoma Region in eastern Tanzania and, within this, the setting of the study area itself. Special attention is given the uniquely tolerant yet predatory relations between chimpanzees and other primates indigenous to the park, particularly olive baboons. Also included as background are descriptions of the research facilities, procedures and techniques insofar as these relate to the observation and interpretation of predatory behavior. A summary of the full range of data gathered on predation serves to focus attention on the episodes observed during the 1968–69 period.

Chapter 2 provides narrative material, in the journal format of the original field notes, on six predatory episodes which have been selected as representative of that behavior. Such explicit accounts rarely appear in research literature, but the episodes are included both for their intrinsic interest and as original information from which insights and conclusions may be drawn by those who could not share in the field experience.

Chapter 3 then draws upon data from all the episodes observed in 1968–69, as well as some of those documented in previous years, to illustrate and discuss predation in terms of three basic stages: the chase, the capture, and the consumption of prey. These activities are described in relation to general behavior patterns exhibited by both predator and prey species, and with occasional emphasis upon such special topics as the importance of control role among adult male chimpanzees during predation, and the learning of predatory habits by subadults.

Chapter 4 consists of a summary and discussion, and is followed by a comprehensive appendix of background information including maps of the study area, tables on the composition of the chimpanzee study community, individual and numerical data on participation in predation, and diagrams of meat distribution. The general purpose of the report is to provide a detailed and complete picture of this behavior in terms of the conditions and circumstances surrounding predation, the activities comprising episodes, and the kind and amount of participation observed.

Because this report is specifically limited to only one aspect of food-getting activity, the procurement of mammalian prey, it is not representative of the full behavioral repertoire of wild chimpanzees. Further information about foraging activities, daily regimen, and social interaction and organization—in general the broad perspective within which predation should be viewed—may be obtained from the publications of J. Goodall or, more recently, J. van Lawick-Goodall. The more immediate objective here is to give a full and accurate description of an unprecedented wealth of information gathered on a primate behavior pattern misunderstood as recently as a decade or so ago. Descriptions of such behavior among one of the closest living relatives of man can perhaps serve also as a fulcrum for raising and weighing hypotheses about the evolution and social development of all primates.

Among many problems inherited by modern anthropologists is the quest for an appropriate descriptive definition of mankind. "It is," as L. Eiseley (1964) so aptly writes, "man's folly, as it is perhaps a sign of his spiritual aspirations, that he is forever scrutinizing and redefining himself." Even after T. Huxley's triumph in 1860, and the resultant surge of interest in man as just another planetary inhabitant, many distinguished individuals were unwilling to consider E. Dubois's suggestions, following his 1891–92 unearthing of pithecanthropus (*Ho-*

mo erectus) remains in Java, that these organisms represented a human ancestral type. Perhaps an aesthetic heritage or an urge to conceive of man as something unique and supreme, a true *Homo sapiens,* was not yet fully erased from the psychological slate, for the basis of this dispute was a widely held exclusive definition for mankind which emphasized the larger size and superior function of his brain.

Twenty years of reconsideration, probably assisted by the importunity of setting meaningful limits to "human" brain volume, resulted in the replacement of this criterion with another: bipedal posture and locomotion. This view was solidly entrenched by 1925, when R. A. Dart initiated new debates about the validity of bipedalism with the unearthing of the first australopithecine (*Australopithecus africanus*) fossils from South Africa. Some years later, when discoveries of pelvic and limb bones indicated erect posture, controversy gradually shifted to a third issue: possession of material culture. More specifically, the new criterion excluded australopithecines from the human line of ancestry on the grounds that stone artefacts were absent from the fossil strata.

Dart responded to these debates with an extensive analysis of faunal assemblages, and by 1955 had produced his osteodontokeratic (bone–tooth–horn) culture hypothesis, which suggested that perishable implements were probably widely used before the advent of stone tools. In spite of considerable evidence pointing to relationships between fractured baboon skulls and antelope bones fashioned into implements, the idea that australopithecines were plains hunters with a material culture gained few adherents in subsequent years because stone tools, which by then were considered more essential to being human than larger brains or bipedalism, were still absent from the sites.

The issue was only partially resolved during the 1960s when L. S. B. Leakey's excavations at Olduvai Gorge, East Africa, produced the pebble tools now known as the Olduwan industry. Although these discoveries and arguments have not yet been resolved, they have continued to encourage the use of a behavioral criterion for distinguishing mankind, the hypothesis being that small communities of "ape-men" probably became more "human" by way of abandoning the forest environment to become skilled tool-making hunters of plains animals.

Another area of interest concerns the world of the cultural anthropologist, particularly in regions inhabited by peoples who practice marginal forms of subsistence. Only slightly removed from the Lake

Tanganyika area are two well-documented examples: the Mbuti Pygmies of the Congo rainforest and the !Kung Bushmen of the Kalahari Desert. The lifeways of such peoples were, early in the century, accepted as representative of a prehistoric stage in human evolution. This view has since altered, and rightly so in some respects. But in line with the new view of stressing the modernity of their humanness, a tendency has developed among ethnologists to avoid behavioral comparison of these human groups with extant primates. The grounds for this reluctance to compare may be enraveled with a human bias to be collectively similar among ourselves while equally collectively dissimilar from other organisms. But the realistic problems are that little behavior can be inferred with certainty from fossil remains, and that the naturalistic behavior of apes and monkeys has been intensively studied only in recent years.

One who has lived with wild chimpanzees cannot but wonder, when viewing J. Marshall's film on Bushman hunting, for example, about the parallels in the ways the two types of primates collect weaver bird eggs and nestlings, gather leaves and berries, use leaves to gather drinking water and to clean the body, or hunt and kill small mammals. Indeed, similarities in actions appear even when the objectives differ: witness a Bushman collecting tubers and a chimpanzee opening a nest to catch safari ants, both using short sticks modified for the purpose.

Yet analogies can be stretched only so far, for these communities, like any others in nature, are alike in some respects and different in others. The Bushman hunts with arrows and spears, the chimpanzee with bare hands; the pygmy usually takes his catch home whereas the chimpanzee does not, and indeed cannot because there is no home base. Clearly there are differences in using a twig stripped of leaves but discarded the same day if not the same hour and in using a sharpened digging stick that may be carried about for months, or in tracking large prey for days at a time versus stalking very young prey for only minutes. The question that always remains, however, is whether these are quantum differences or just variations on basic behavioral themes.

Considering some of these problems, C. R. Carpenter, H. W. Nissen, and others initiated systematic field studies of nonhuman primates in the early 1930s and thereby brought gradual attention to another source of information for students of primate evolution and human origins. By the 1960s numerous major field studies of monkeys and apes were completed and in progress. Prime examples are the work of S. L. Washburn, I. DeVore and K. R. L. Hall with savanna baboons in East and South Africa, of G. Schaller and D. Fossey with the mountain gorilla in Central and East Africa, of J. Goodall with the chimpanzee in Tanzania, and of numerous Japanese primatologists—K. Imanishi, J. Itani, K. Izawa, M. Kawai, A. Kawamura, T. Nishida, Y. Sugiyama, A. Suzuki, and others—with macaques or langurs in Asia, and with chimpanzees in East Africa.

The discoveries of J. Goodall in particular, among which were the first detailed descriptions of naturalistic implement manipulation and production as well as hunting and meat-eating among chimpanzees, are major contributions to the understanding of primate behavior and evolution. More recently, the findings at the Gombe Stream Research Centre have been greatly elaborated and refined by additional years of field observation, years which are in many ways only beginning to yield the results necessary for understanding the subtle behavioral systems that are exhibited by chimpanzees, and the results that are at the same time pertinent to the study of human origins and development. To date, these results include multifaceted social relations and organization strongly influenced by personality, lifetime mother-offspring and sibling relationships, preferential sexual pairing that coexists with group sharing of estrous females, incipient avoidance of intrafamily sexual relations, infrequent aggressive competition for preferred foods and females, complicated vocal and gestural communicatory systems, varied and dextrous use of objects, and omnivorous dietary habits.

The present report focuses on this last subject. Ranging far beyond the point of recognizing an element of occasional, opportunistic predation among these apes, many aspects of predatory behavior will be discussed: the repeated attempts to capture prey of varied species; the coordination and cooperation involved in group stalking of prey; the voluntary sharing of meat along lines not always determined by social rank; the patterned technique of breaking into skulls; and the "division of labor" implied by mainly adult males participating in chases and captures. Presumably it is the conditions within which and the modes whereby predation currently operates among wild chimpanzees that may yield additional clues to our mutual prehistory.

The manner in which field observations are made and data are eventually handled is likely to be much influenced by the aims, training, and experiences of the person (s) concerned. During my first weeks at Gombe I was bewildered, and at times disheartened, by the volume of details and skills which had to be learned, from the previous experiences of my associates to the names and other distinctive features of nearly fifty highly individualistic chimpanzees. However, the ensuing months brought a degree of familiarity which at times tended to border on complacency, for I began to feel secure in my newfound grasp of the many components which constitute and regulate chimpanzee society and activity. Had I left the Gombe at this point, my willingness to discuss the generalities of chimpanzee behavior would have been relatively unfettered, and my knowledge both shallow and naive. Only after about eight months did this assurance begin to dissipate as I became more deeply familiar with each individual as a composite of distinct movements, interaction styles, mannerisms, preferences, and other qualities. My eagerness to discuss general patterns in the chimpanzee mode of life rapidly decreased during this stage of the field experience, and I discovered that my new response, when asked, let us say, about the display activities of adult males, was to consider that Mike does this one way, Hugo another, and Hugh very differently from either. Only after many more months passed did these variations begin to coalesce into an understanding of composite behavioral patterns. These developments later led me to speculate that any field study which does not allow sufficient time for the observer to pass through a series of learning stages runs several risks, the foremost of which might be a premature sense of satisfaction about recognizing and understanding the broader aspects of behavior and social organization. These problems are faced but not necessarily surmounted in the ensuing text, and the reader should allow for their effects.

The approach used in presenting and analyzing numerous predatory episodes is also influenced by my own academic training in anthropology and by the current goals of the research center. A general outline of chimpanzee behavior within the Gombe study community was blocked out some years ago, forming an accurate but abstract image of the way in which we then conceived of wild chimpanzees. Although information continues to accumulate along many broad lines at Gombe, resulting in periodic revisions to basic concepts, the study emphasis in recent years has increasingly been upon the relevance and importance of individualistic behavior and personality development within frameworks of social interaction and organization. The knowledge derived from these goals—much of which is informally transferred to any researcher working a year or more as a team member—can lead down unending paths of conceptual associations even when the study is a clearly defined one of predatory behavior. Thus, emphasis upon individualistic behavior, availability of an abundance of information peripheral to any special study, and an anthropological perspective are all reflected in this report on chimpanzee predation.

THE PREDATORY BEHAVIOR OF WILD CHIMPANZEES

"They [chimpanzees] subsist on wild fruits of various kinds, but they will also visit forsaken plantations, and even those which are still under cultivation, and in some cases it seems that they do not reject animal food."

R. Hartmann (1886:237)
By courtesy of Appleton-Century-Crofts

"But they [chimpanzees] are known to be very adaptable in diet, and one who ventures to describe them as vegetarians is sure to be given the lie humiliatingly by an individual who happens to prefer meats to vegetables."

R. M. Yerkes (1943:15)
By courtesy of Yale University Press

"Hunting [by man] not only necessitated new activities and new kinds of cooperation but changed the role of the adult male in the group. Among vegetarian primates, adult males do not share food. They take the best places for feeding and may even take food from less dominant animals."

S. L. Washburn and V. Avis (1958:433)
By courtesy of Yale University Press

"I suggest that meat-eating is as old as man the tool-maker, that with adaptation to partly open forest margins the diet of proto-men inevitably became more varied, and that they changed from being eaters largely of plants and the fruits of plants to being in part meat-eaters."

K. P. Oakley (1961:189)
By courtesy of Aldine-Atherton Inc.

"Given the division of labor by sex and the formation of domestic units through marriage, it follows that sharing food and other items, rather than being nonexistent, as among monkeys and apes, is a *sine qua non* of the human condition. Food sharing is an outstanding functional criterion of man."

M. D. Sahlins (1965:66)
By courtesy of Wayne State University Press

". . . we must not start running away with the idea that chimpanzees are 'primitive hunters.' They are vegetarians. The meat-eating incidents mentioned above are extremely rare in the apes studied by Jane Goodall, and her apes were unusual and atypical of the species in general, living as they do in un-chimplike surroundings."

R. Morris and D. Morris (1966:228)
By courtesy of McGraw-Hill Book Co.

Mike (MK)
prime adult
(alpha male)

Goliath (GOL)
past-prime adult
(ex-alpha male)

Leakey (LK)
past-prime adult

Hugo (HG)
past-prime adult

Hugh (HH)
prime adult

Humphrey (HM)
prime adult

"PREDATORS"

Worzle (WZ)
prime adult
(sickly)

Faben (FB)
prime adult
(polio paralysis of right arm)

Pepe (PP)
prime adult
(polio paralysis of left arm)

Charlie (CH)
young adult

Evered (EV)
young adult

Figan (FG)
young adult

1

DISTRIBUTION AND FIELD STUDY OF CHIMPANZEES

Chimpanzees of various species[1] are distributed throughout portions of West, Central, and East Africa, with approximate maximum range extensions to latitudes 13°N and 8°S (Kortlandt and van Zon, 1969).

The exact East African distribution of *Pan troglodytes schweinfurthii,* the eastern or long-haired chimpanzee, is not known. But these great apes are for the most part confined to a 50-mile-wide belt running from northwestern Uganda, near the northern tip of Lake Albert, to southwestern Tanzania, almost to the southern end of Lake Tanganyika. This rain forest and savanna-woodland belt follows the western trough of the African rift system. The best documented localities within the belt are in western Uganda and Tanzania. Kenya contains no indigenous populations; Rwanda and Burundi presumably do, for each country intersects the major zone of habitation, but no recent survey of these regions has been made. Long-haired chimpanzees are also found in parts of Central Africa (Congo), rough boundaries being the Ubangi River to the north and the Lualaba River to the west.

All known habitats of the long-haired chimpanzee in Uganda are semi-isolated, forested areas adjacent to the Congo border, from Murchison Falls National Park at the north to the Rwanda border at the south (Reynolds and Reynolds, 1965). A similar distribution prevails in Tanzania, where isolated or semi-isolated populations are scattered along the eastern shore of Lake Tanganyika. Reynolds

1. Taxonomists disagree as to the appropriate specific and subspecific nomenclature for the genus *Pan*. For present purposes it is sufficient to differentiate the Eastern variety (*P. troglodytes schweinfurthii*) from the Central (*P. t. troglodytes*) and Western (*P. t. verus*) varieties—all of which are sometimes known as *P. satyrus*—as well as from the Pygmy (*P. paniscus*) variety. (See Napier and Napier, 1967, for details.)

(1963, 1964, 1965a–c) conducted an 8-month field study of chimpanzees in the Budongo Forest of Uganda in 1962, Sugiyama (1967, 1968, 1969) a 6-month study in the same area in 1966–67, and a third study by A. Suzuki may still be in progress. Two long-term studies have been initiated in Tanzania, one in 1960 at the Gombe National Park (Goodall, 1962, 1963a–b, 1964, 1965; Lawick-Goodall, 1965, 1967a–b, 1968a–b) and another in 1961 in the Kasakati Basin area by Japanese primatologists (Azuma and Toyoshima, 1962, 1965; Itani, 1965, 1966, 1967; Itani and Suzuki, 1967; Izawa, 1970; Izawa and Itani, 1966; Kawabe, 1966; Nishida, 1967a–b, 1968, 1970; Suzuki, 1966, 1969). In addition to these studies in East Africa, numerous primatologists have conducted survey expeditions and studies in West and Central Africa during the past decade (Jones and Pi Sabater, 1969, 1971; Kortlandt, 1962, 1963, 1964, 1965a, 1966, 1967, 1968a–b; Zon and Orshoven, 1967).

THE GOMBE NATIONAL PARK OF TANZANIA

The Kigoma Region of the United Republic of Tanzania, East Africa, comprises approximately 17,000 square miles of differentiated terrain bounded by Lake Tanganyika to the west, the Burundi nation and Kasulu Region to the north, and the regions of Kibondo, Tabora, and Mpanda to the east and south. Kigoma Region supports a primarily agricultural-fishing population of nearly 400,000 from mainly the Ha, Rundi, Nyamwezi, Binza, and Kiko tribes. The port town of Kigoma (pop. ca. 16,000 including Ujiji) serves as a center for governmental administration and trade, and is accessible by boat across the lake, a single cross-country rail line through Tabora from the east, three unpaved roads from the north, east, and south, or by chartered airplane. Distances from East African towns of first importance, such as Kampala in Uganda, Nairobi in Kenya, and Dar es Salaam in Tanzania, are from 600 to 800 miles. The rail trip from the Indian Ocean coast, for example, requires 3 days, but surface travel is often restricted to the dry season because of flooding of roads and rail lines. (Appendix A, Map I.)

The Gombe Stream Game Reserve was established in 1945 as a conservational measure with the aim of preserving the indigenous chimpanzee population. This area, which is small relative to most East African game parks, officially became the Gombe National Park in 1968. The park is located on the eastern shore of Lake Tanganyika, extending some 10 miles between the Burundi border and the town of Kigoma. Geographic and agricultural boundaries enclose the park area—the lake and rift scarp to the west and east, settlements to the north and south—and access is limited to foot paths and boats due to the ruggedness of the rift scarp, which rises to 2500 feet above the lake level. (Appendix A, Map II.)

PHYSIOGRAPHY, CLIMATE, AND WILDLIFE

The physiography of Kigoma Region is diversified in comparison to some of the central regions of Tanzania. The highland plateau of East Africa rises to form a series of fault scarps flanking the western trough of the African rift system, occupied here by the 420-mile-long and at least 1-mile-deep Lake Tanganyika. East of the scarp crest, at a point about 10 miles from Kigoma, lies the extensive Malagarasi drainage basin, which floods seasonally into lakes and swampland. Extending north and south is an irregular scarp line that parallels the lake shore at 2 to 3 miles distance, with a region of lowlands south of Kigoma where the Malagarasi River cuts through to the lake. Between the scarp crest and the lake shore, a distance of about 3 miles, are a series of erosion-dissected ridges. Across the 35-mile width of the lake are a line of similar highlands along the Congo shore.

The Gombe National Park (Plate 1), which begins about 6 miles north of Kigoma harbor and extends some 10 miles along the lakeshore and 2 or 3 miles inland to the rift scarp, includes an area of about 25–30 square miles.[2] The terrain in the park is rugged, with gradients up to 1000 feet per mile between the escarpment and lake, whose surface is more than 2530 feet above mean sea level; slopes on the ridges separating main valleys are even steeper, often between 45° and 60° from the horizontal. These main valleys all trend from east to west into the lake, the outlets intersecting the beach from 1 to 2 miles apart. The predominant drainage pattern is dendritic, controlled mostly by the steep slopes. Main valleys from the scarp crest are met by ravines trending north–south from the dividing ridges. Most large valleys contain perennial streams fed by springs near the steep upper slopes of the rift scarp; but some valleys and most ravines have running water only during the rainy seasons. (Appendix A, Map III.)

In general, Kigoma Region has a tropical savanna climate charac-

2. These projected map areas are not equivalent to true surface areas, the latter being perhaps as much as twice the size of the former due to the very irregular terrain.

PLATE 1

(Left) View from Troop Knob along Mkenke Valley, showing marked vegetation differences between the lower ridge, the valley bottom, and the high escarpment. (Top right) A section of Gombe National Park showing the rugged terrain between the lake and rift escarpment, early rainy season. (Bottom right) Northward view across the ridges separating the lower portions of Mkenke, Kakombe, and Kasakela valleys, late dry season.

Eastward view along Kakombe Valley, with Peak Ridge on left, Sleeping Buffalo Ridge on right, and the rift escarpment ca. 2.5 miles from the beach.

Rainy and dry season views of the feeding area in Kakombe Valley.

terized by lack of frost, a mean annual rainfall of 30–50 inches, and a diurnal temperature range of about 40°F. Within the park, the average diurnal temperature range is about 15°F, daily ranges showing only slight seasonal variation. Temperatures may drop to 62°F at night and rise to 98°F at midday. The rainfall at the research center was 64.7 inches during the study period, between March 1968 and March 1969. The rainy season usually runs from October to May or June and this period divides almost equally into two phases: the short and long rains. Brief intervals of daily precipitation introduce the rainy season, but the cloud cover is usually so scattered that soil and vegetation dry regularly. Within a few weeks, however, the yellows of dust and dead plants are replaced by green vistas of growing vegetation. The sporadic rain pattern changes around February to steady rains which may persist for days at a time, with only heavier showers breaking the monotony of nearly constant precipitation.

The last few months of rain affects both work and morale because the valley bottom vegetation becomes impenetrable, grass on the higher slopes may reach heights of 14 feet and more, and people and equipment are constantly damp.

The first easterly winds introducing the dry season are welcomed, for the skies clear and surfaces dry within a week or two. During subsequent months the vegetation changes as grass and underbrush die out in the valley or burn away from the high slopes of the scarps. Though skies remain very clear for months, visibility is actually reduced during the dry season because of the dust and ash of grass fires transported by strong easterly winds.

The ecology of this portion of East Africa appears complex but remains relatively unknown. Kigoma Region supports a predominantly Miombo woodland which is interspersed by Acacia savanna in the Malagarasi and Highland portions. There are both open grasslands and typically dense tropical forest containing representative flora and fauna. The Gombe National Park resembles this pattern in miniature, exhibiting a wide range of intersecting zones and niches compressed into a few square miles (Plate 2). Grasses, vines, shrubs, and trees occur in great variety. Vegetation distribution in the park seems determined for the most part by topography. Valley bottoms and lower ridge slopes are typically of the riverine forest variety found extensively along the lake, and an eastward walk upslope beside a main stream can include the open floor of gallery forest as well as pockets of densely interwoven thickets.[3] A turn north or south toward

3. Soon after my arrival in March, when still a novice as to efficient routes and trails, in one valley I made a 6-hour excursion that covered less than 2 miles.

PLATE 2

Contrasting vegetation cover: open grassland on the western slopes of Sleeping Buffalo Ridge and the forested floor of lower Kakombe Valley.

The woodland-grassland transition on the lower slopes of Sleeping Buffalo Ridge, in Kakombe Valley.

The relatively open floor beneath forest canopy in Kakombe Valley, near the base of Dung Hill.

View along a primate trail, passing through the typically dense thicket of a ravine, or korongo (KK 10).

one of the main ridges may then begin a climb into a heavily overgrown ravine (korongo) that affords only brief glimpses of the sky, or through the more open Brachystegia woodland and grassland of the hillsides. Achieving the ridge crest, another turn eastward leads to the tree line, perhaps 1700 feet above the beach, and past this onto the grassy, boulder- and cliff-strewn slopes of the scarp crest itself.

The park fauna at first appear less profuse than the flora. This is due in some part to the relative paucity of large mammals that are so common to other parks in East Africa. Observed exceptions to this rule include an occasional leopard and numerous African buffalo, as well as bushbuck, bush pig, and monitor lizard. Small mammals, reptiles, and birds abound, however, and such nocturnal species as the African civet, the palm civet, the large-spotted genet, and the serval are regularly observed. Primates other than the chimpanzee include olive baboons (*Papio anubis*), red colobus (*Colobus badius*), redtail (*Cercopithecus nictitans*), vervet (*Cercopithecus aethiops*), and silver and blue monkeys (*Cercopithecus mitis*), as well as the galago (*Galago spp.*).[4]

Knowledge of the park ecosystem is still very incomplete even though many species of flora and fauna have been tabulated during the past decade. Little has been systematically studied beyond those problems pertinent to observation of chimpanzees, baboons and colobus monkeys. However, even limited evidence yields some general impressions, one being that the narrow zone between the lake shore and rift scarp harbors a network of flora and fauna that differs from contiguous regions east of these highlands and across the lake. In part this difference may be a result of extensive human settlement of the area between Lake Tanganyika and the Malagarasi River floodplain. At any rate, many species living within the park appear to be absent or rare beyond the rift highlands, and some possibility exists that this narrow zone is a blend of East and Central African ecosystems. The steep rift scarps and rugged terrain along the lake shore may constitute a partial barrier between the Central and East African areas, creating thereby a special ecosystem that has an abundance of nonhuman primates. The limited distribution of chimpanzees along the Tanzanian border of the lake may be related to these phenomena.[5]

4. Relations between chimpanzees and other primates, as well as other mammals, reptiles, amphibians, birds, and insects have been described by Lawick-Goodall (1968b: 173–6, 291–4).

5. Background sources for this section include: Holmes (1965), Maberly (1965), Trewartha (1954), Williams (1968), the Shell Road Map of East Africa (1965), the International Map of the World series 1301 (1964), the Governmental Survey Map of Tanzania (1965), the Geological Survey Map of Kigoma Quarter Degree Sheet 92 (1961).

RESEARCH FACILITIES AND PROCEDURES

The Gombe Stream Research Centre is located in the lower portion of Kakombe Valley (Appendix A, Map IV), some 12 miles north of Kigoma harbor. The main field station at Kakombe is divided into two sections, one built at the beach near the mouth of the stream and the other about half a mile inland, just above the valley floor on the slope of Peak Ridge. Thatched, prefabricated aluminum buildings of varying sizes house research personnel, supplies, and equipment. The lower station, or Beach Camp, is the logistics base for all types of research projects and consists of a number of storage and residential units, a main kitchen, photographic darkroom, and boathouse. A steep main path, paralleling the stream course, leads past additional housing units and a storage building for the bananas used in artificial feeding. This path leads to the upper station, or main camp, which is the residential and research base for the chimpanzee study program (Appendix A, Maps IV and V). Located perhaps 60 yards upslope from the stream, this camp is dominated by a large central building surrounded by 40 concrete-encased metal feeding boxes that are evenly dispersed over a circular area of about 1200 square yards. Underground electric wires lead from a battery to the solenoid latches of these boxes, so that each one can be opened from indoors. A 40–foot underground trench, the latest in a long series of attempts to improve feeding techniques (but not yet in use during 1968–69), leads from an access door in the building to similar feeding boxes. Beyond the periphery of the feeding area, which is trimmed weekly during the rainy season, are additional housing units for researchers working with chimpanzees (Plate 3).

Six primate trails, as well traveled now by humans as by baboons and chimpanzees, radiate from the feeding area. The three main valleys of the central study area—Kasakela to the north, Kakombe, and Mkenke to the south—can thus be reached by experienced personnel within an hour.

Data recording equipment includes items used by all investigators as well as items assigned to the use and care of individuals. A systematic photographic record begun by H. van Lawick in 1962 is maintained on subjects of both general and special interest. Selected staff members receive training with an Arriflex 16mm cine camera, but the scientific record is supplemented mostly by greytone and color still photographs and Kodachrome Super–8 cine. Nikon 35mm still cameras (with motor drive, and lenses from 50mm to 200mm) and a

PLATE 3

Two views of the feeding area, showing the central research building and the concrete-encased feeding units distributed around it.

With banana feeding in progress, the dispersal of chimpanzees on the boxes illustrates the utility of this design in keeping aggressive competition to a minimum.

Nikon Super–8 cine camera, always kept on hand in the central building, are frequently used to document predation. Because observation distances have been considerably reduced over the years as chimpanzees became neutrally conditioned to researchers, most photographic work can easily be done with lenses in the 50–135mm range. The pictures in this report, for example, were taken by the author with 50–58mm lenses on manual Nikon and Exacta still cameras.

Several portable radios of the walkie-talkie type are available for use in special work or emergencies, and a portable Nagra sound recorder with standard 5-inch reels is used for recording chimpanzee vocalizations. Each researcher is provided with items such as a medical safety kit (containing snake antivenin and other necessities), a portable Phillips cassette recorder, and a typewriter for transcribing daily notes.

A general policy to which all staff members conform has required, since 1963, complete noninterference and noninteraction by researchers in regard to chimpanzees. But any field study of naturalistic behavior, and especially of a species inhabiting such rugged terrain, must at some level compromise human intervention with information retrieval. Feeding chimpanzees in order to condition them to neutral observers, thereby increasing the quantity and quality of certain types of data, is one such compromise, and one applied at Gombe with care and surveillance. Bananas rarely grow wild in the habitat of these chimpanzees, so the introduction of this favored fruit (in 1962) has facilitated close observation of individuals and increased the consistency with which some kinds of data have been collected. Regular feeding was originally attempted from a variety of containers, but, in the words of Lawick-Goodall (1967a:44): "Every type we tried the chimpanzees opened either by brute strength or manipulative cunning."

The pattern for ensuing years was thus set early, inasmuch as researchers have periodically been forced to redesign the feeding system in order to maintain control over distribution and competition. The first technique consisted of concrete containers set into shallow pits. These the chimpanzees soon pried open by hand or with sticks. More complicated boxes with metal lids fastened by wires were installed a few years later in a new feeding area at the present site of main camp. Wires ran through underground pipes to the central building, but the chimpanzees soon learned to pull the fastening pins, and when nuts and bolts were substituted these also were removed. By 1967, when the study community had increased to nearly 50 individuals because of regular feeding, new electrical boxes were installed. For some months this innovation seemed effective and in keeping with the objective of feeding to provide maximum observability while stimulating as little abnormal behavior as possible (Lawick-Goodall, 1968b).

Hand feeding of bananas was eliminated—boxes were filled at night and opened remotely—so that humans would not be associated with this highly preferred food. There were enough boxes containing a sufficient supply of bananas to feed the entire study community at one time without stimulating unusual competition, and the spacing and control of boxes effectively dispersed individuals over a large ground area; both factors minimized abnormal behavior. Feeding occurred almost daily during 1967, but this was reduced to a random 7 days per 2 weeks in early 1968. This decrease was the first step in returning to more natural conditions, a reversal made possible (without a corresponding decrease in the range or consistency of data collection) by the high degree to which chimpanzees had accepted humans in all locations and circumstances.

During 1967 one of the local baboon troops, the range of which has always included the feeding area, began to visit main camp very regularly during morning feeding hours. Rewarded by the availability of discarded bananas and peels, these baboons rapidly accommodated themselves to the presence of humans and chimpanzees in large numbers. There soon arose problems with interspecific aggression, and all phases of work at the feeding area were hampered by so many primates (sometimes as many as 100) concentrating in such a small area. By June 1968 reorganization of feeding procedures was again necessary. Feeding was reduced to once or twice weekly, sometimes less, and unused bananas and peels were collected immediately. The daily movements of baboons were kept under close surveillance, and feeding was planned in accordance with their location in Kakombe Valley. Data-collection methods were also changed, so that observation was less dependent upon chimpanzees visiting camp for long periods each day. The return to nearly natural conditions was completed by early 1969, when bananas were distributed through an underground trench from which the fruit could be *inserted* and *removed* at will without visual exposure to chimpanzees or baboons. Only individuals or small groups were fed, and each chimpanzee received a few bananas about once every 2 weeks.

As daily activities are observed, the data are voice-recorded on cassette tapes; the taped data, most of which are timed to the minute, are transcribed daily to a typewritten, running record of collated information. This journal is supplemented by various charting and filing systems, including photography. Data collection is currently carried out by researchers who are able to move about freely among most members of the study community, at distances of only a few feet even when chimpanzees are far away from the feeding area, because neutrality is strictly maintained.

Because of the irregular and unpredictable occurrence of predatory behavior among the chimpanzees, no special techniques or methods have been applied to its documentation. The feeding conditions, recording techniques, and observation methods described above for general research apply to predatory behavior as well. However, predation does receive maximum coverage when it occurs, for all available personnel usually assist. The common procedure may be outlined as: (1) the first person on location starts recording the central activity around the prey; (2) if the main group of chimpanzees divides into smaller units, another person may observe peripheral individuals and clusters; (3) should the episode last many hours, standby researchers may spell the main observer, and one or more persons may take still and motion pictures; (4) all recordings, independent observations, and films are collated for the final notes, preferably that same day.

Documentation of predatory behavior was often supplemented during the 1968–69 period with relevant data supplied by persons studying the local baboon troops. Most predatory incidents lasted only a few minutes in cases where capture failed, so it frequently happened that only one person saw the sequence; on the other hand, kills usually lasted several hours and were attended by at least 2, and often 4 or 5, researchers working in various capacities. Only 2 of these, usually the main observer and photographer, worked at close quarters to the activity. Normal observing distances during predation were from 1 to 5 yards on the ground and up to 20 yards (with Leitz binoculars) when the chimpanzees took the meat into trees (Plate 4).

THREE PRIMATE STUDY COMMUNITIES

A study of predation should approach this behavior from two viewpoints: that of the predator and that of the prey. As in many field studies where predator and prey are not equally habituated, this was

Patrick McGinnis (left) and David Bygott photograph a meat-eating cluster of adult chimpanzees.

(Left) Patrick McGinnis observing predatory behavior at close quarters. (Right) A sample result of such close photography, showing the adult chimpanzees Mike (holding skull), Leakey (biting ribcage), and Flo begging from Hugo as they cluster around a freshly killed colobus carcass.

not always possible at Gombe. But the situation during 1968–69 was favorable for this dual approach because most of the prey were baboons and because the baboon troops concerned were under regular observation. As a prelude to discussing predation, it is useful to briefly describe the 3 primate study communities. The baboon communities are natural units, or troops, which are essentially cohesive and stable in time and location. Although there exist some hypotheses concerning the social units of chimpanzees (Itani and Suzuki, 1967; Izawa, 1970; Lawick-Goodall, 1968b; Nishida, 1968; Reynolds and Reynolds, 1965), the structure of chimpanzee society remains in doubt. Because there is no certainty that the fluid groups regularly observed among Gombe chimpanzees form a finite, cohesive community, those chimpanzees under observation can at best be considered a *study community.*[6]

The *total chimpanzee population* of Gombe National Park has not yet been determined. A numerical estimation would be in the 100 to 150 range (Lawick-Goodall, 1968b). Assuming that banana feeding has not excessively distorted distribution within the 25–30 square mile park area, the population may be as high as 200 to 250 individuals. The size of the habituated study community—that is, the known individuals regularly visiting the camp area—varied in recent years from 40 to 50 members. The 1968–69 predation data were collected from a study community comprising 48 frequently seen individuals, of which 6 had died or were missing by mid-1969 (Appendix B, Tables I and II). Of these 48 chimpanzees, 13 adult males were most often directly involved in predatory behavior.[7] The *total baboon population* is also undetermined. However, basing calculations on the ranges and sizes of 2 troops regularly observed in Kakombe Valley, a rough estimate for the park region would be about 25 to 30 troops, or about 1000 to 1500 individuals (Ransom and Ransom, in press).

Group size within the chimpanzee study community varies greatly: the most common groups contain from 2 to 6 individuals, but temporary associations can range from 2 to more than 30 (Lawick-Goodall, 1965, 1968b).[8] The higher numbers are most rare, and large feeding

6. A detailed study of ranging habits, daily activities, and social organization was initiated by the author in August 1970.

7. Because names are a useful shorthand in a study that focuses upon individuality, these are used in all specific contexts not requiring age-sex classification. Letter abbreviations for the 13 adult male names appear in tables, figures, etc.

8. *Group* being here defined as a temporary association of individuals remaining constant for hours or days (Lawick-Goodall, 1968b). *Cluster,* as used later in the text, refers to similarly temporary subunits of mutually preoccupied individuals within the larger group context.

groups appear more common than large travel groups. This is particularly true in predatory episodes where the average size of groups is much greater than 6. In early 1968, when bananas were available nearly every other day, attendance at camp frequently exceeded 35 chimpanzees at one time, a number seldom encountered as a unit in other situations. The feeding changes introduced in June of that year gradually reduced this high attendance rate since more staggered and limited feeding resulted in more normal aggregations of 2 to 10 individuals. Most members of the study community ranged throughout the approximately 10 square miles of the main study area, and some consistently traveled far greater distances. Individual and group ranges have not been sufficiently studied to provide accurate data on range limits or preferred core areas. Mature chimpanzees probably have a range of 15 to 20 square miles and some individuals may range throughout the 25 to 30 square miles in the park; mothers with infants seem to have smaller ranges of 8 to 10 square miles.

In the 28 observed cases where baboons were preyed upon by chimpanzees during 1968–69, the incidents involved the local baboon communities known as Camp Troop and Beach Troop. The former consisted of about 75 individuals,[9] but the troop often fragmented into smaller groups which sometimes remained apart overnight. Also, lone adult males and small mixed groups were occasionally encountered during daylight hours at considerable distances from the main body of baboons. By early 1968, Camp Troop was rapidly becoming conditioned to humans and was increasingly drawn to the feeding area on feeding days by chimpanzee activities during banana distribution. Beach Troop, which contained about 55 individuals, visited the feeding area less frequently. They attended some banana days when Camp Troop was elsewhere, but rarely entered main camp or stayed a long time. A few adult males would sometimes appear without the rest of the troop, and these would occasionally mingle with Camp Troop baboons during feeding. The ranges of both troops included large inland and waterfront segments (Appendix A, Map III), and the 2 ranges overlapped to some extent in Kakombe Valley. Daily Camp Troop movement centered on a core zone comprising the lower floodplain of Kakombe Stream, Mango Ridge, and the southward extension of the beach to Mkenke Valley, with 2 sleeping sites at the Dell and the mouth of Mkenke Stream. The core zone of Beach Troop extended roughly from the mouth of Kakombe

9. Absolute numbers in both troops varied considerably during the year due to the fluctuation of birth and death rates, the latter including predation by chimpanzees and killing by local fishermen.

Stream across Peak Ridge to Hidden Clearing, down along Kasakela Stream, and back again along the waterfront, with one sleeping site at Hidden Clearing.

A third and possibly fourth troop occupied the upper portions of the three main valleys in the study area. Because these were not at all habituated to human beings, chance encounters while following chimpanzees usually resulted in alarm vocalizations followed by swift departure. A few unidentified males who occasionally appeared singly or in pairs at the feeding area were probably from these higher areas.

AGONISTIC, TOLERANT AND SOCIAL RELATIONS AMONG STUDY COMMUNITIES

Relations and communication (vocal and gestural) between chimpanzees and other primate species in Gombe National Park, particularly baboons, have been described to some extent by Lawick-Goodall (1968b:294), who concludes that these are "highly variable, ranging from complete tolerance to violent aggression, hunting, and killing," and consist of "a variety of subtle communicatory signals such as nonsexual presenting, mild threat, and invitations to play and groom." Here it is necessary to discuss only briefly some of the common social relations forming a larger behavioral backdrop wherein predation occurs as a sporadic, or part-time, activity initiated only by chimpanzees intending to capture and consume young baboons.

Chimpanzees and baboons in Gombe National Park have regularly been observed to be tolerant of one another even in natural feeding conditions, or to show mutual avoidance within overlapping ranges. A similar statement can be made about relations between chimpanzees and other, smaller primate species in the park. Young chimpanzees sometimes play with monkeys, and one subadult has been observed in relaxed contact play with an adult redtail for half an hour. Only a very small percentage of encounters between monkeys and chimpanzees lead to predation. "This clearly demonstrates," according to Lawick-Goodall (1968b:191), "the danger of drawing conclusions about the predatory tendencies of a species on the basis of a few such negative encounters (see, for example, Reynolds and Reynolds, 1965; Kortlandt, 1965)."

Notable exceptions to tolerance are the very rare observations of small groups of chimpanzees being harassed or chased by baboons or colobus monkeys even when there were no visible indications of predatory intent. Beyond such glancing generalizations, interspecific relations are difficult to comprehend or describe precisely because of great variability in conditions and circumstances. For example, the normally utilized foods and preferred areas in the park differed in many respects even for terrestrial baboons and chimpanzees. Both Beach Troop and Camp Troop baboons often foraged along the waterfront and even competed with local fishermen for daga, a sardine-like fish traditionally dried in the sun. Although chimpanzees sometimes visited fig trees or berry bushes near the outlets of major streams or on Beach Ridge, they rarely wandered onto the open beach and never did so when unknown Africans were present. Similarly, ground foraging for a large variety of microedibles was common with baboons and rare among chimpanzees. Nonetheless, the 2 species were frequently within hearing distance, were often mutually visible, and were sometimes completely intermingled. Mutual avoidance was more common than interspecific aggression in these natural circumstances.

Banana feeding clearly affected some interspecific relations during 1968–69. Baboons and chimpanzees spent more hours per day together in larger than normal concentrations while at the feeding area, and agonistic behavior (i.e. threats, harassment, attacks) increased during banana distribution. Yet for every case of predation possibly connected with these circumstances there were many other times when the 2 species coexisted amicably in the same limited space. Chimpanzees numbering in the 30s and baboons in the 50s, many within a few feet of each other, would remain for hours in the vicinity of camp without being aggressive toward each other. On days when bananas were available, the most intense aggression between species accompanied the start of food distribution or the unexpected arrival of many baboons when boxes had just been opened. There would be a rapid decrease of agonistic behavior immediately following the initial excitement and competition over bananas. Then, when feeding was discontinued or completed, tolerance would again prevail. Similar tolerance was observed occasionally at the feeding area among members of different troops. Adult males from Beach Troop were accepted by Camp Troop, and some baboons were even known to change troops without causing major conflicts. Thus, tolerance between chimpanzees and baboons has been and continues to be the most common relation regardless of location. Generally speaking, this state of coexistence held in 1968–69 no matter how many attempts or kills were

As Leakey (left) eats bananas brought from a feeding box, a male baboon gradually inches toward the fruit. Just before touching the bananas, the baboon suddenly pulls back as Leakey threatens him.

Holding several bananas that male baboons also want, Worzle first "pant-hoots," then threatens by waving his arm, and finally hurls a stone at them.

PLATE 6

(Left) Chimpanzees eat bananas, then discard peels, which are collected by baboons. (Right) After feeding, both chimpanzees and baboons relax and intermingle tolerantly throughout the feeding area.

A typical play cluster of two young baboons and two infant chimpanzees, with a young adult male baboon and the mother of one chimpanzee infant sitting at the periphery of the play area.

An example of idle "sparring" play between an infant chimpanzee and a juvenile baboon.

made per month and despite the fact that in nearly all banana feeding contexts, even with the preventive measures taken by researchers,[10] agonistic interactions were more frequent and intense than in the natural conditions away from the feeding area.

During banana distribution most conflicts were resolved by retreat or mild threat, with relatively few cases of violent bodily contact such as grappling or biting (Plate 5). And in the latter case, attacks were always brief, terminating when active defense occurred or when the bananas were acquired and taken away or eaten. Minor injuries such as slight scratches or torn-away hair were occasionally recorded, but no competing chimpanzee or baboon was seriously disabled during the year, even in the interspecific affrays that sometimes accompanied predation.[11] Contrary to the flight or retaliation common to intense competition, some agonistic situations resulted in submissive responses by baboons toward chimpanzees. Appeasement behavior, including such baboon gestures as presenting and foot-backing (approaching backwards and raising a foot to touch another's body), was occasionally seen among subadults of these species.

A variety of nonaggressive interactions regularly occurred at the feeding area throughout the 1968–69 period (Plate 6). Interspecific behavior perhaps best exemplifying nonaggressive relations involved grooming and play. Although grooming was rarely observed, play between juvenile baboons and infant chimpanzees was quite frequent, happening weekly and sometimes repeatedly in one morning or afternoon. Neither activity was likely to occur during banana distribution and consumption, coming instead during the period of relaxed activity which followed or on nonfeeding days when the 2 species happened to meet in the area. The play invitations of young originated from individuals of either species and, when accepted, the ensuing interactions would usually consist of touching and light slapping, pulling and pushing, or chasing. Prolonged wrestling of the type often seen among infant chimpanzees was not often observed, perhaps because of its rarity among baboons. Play clusters sometimes contained 2 or 3 members from each species, though a single pair was most common. Mothers interfered with play only in case of violence or

10. Preventive measures included stopping feeding, closing boxes, collecting bananas and peels, standing near chimpanzees, and threatening some of the more persistent baboons.

11. Considering the efficiency of defensive or protective aggression among savanna baboon troops (DeVore and Hall, 1965; Washburn and DeVore, 1961a–b) and the manner in which a serval cat was dispatched by Gombe baboons in 1968, the absence of serious injury to predator chimpanzees seems remarkable.

distress, and, unlike male baboons, male chimpanzees tended to ignore these disturbances.

Perhaps the most intriguing aspect of chimpanzee and baboon coexistence at Gombe is the mixture of tolerant and predatory activity and the rapidity with which the former would sometimes become the latter. In several instances at or near the feeding area, predatory behavior would suddenly erupt from prolonged, relaxed situations in which members of both species had been ignoring one another or even amicably interacting. Once a juvenile baboon who had been playing with an infant champanzee was captured and eaten only hours afterwards by adult chimpanzees.

PRIMATES AS PREY

Nonhuman primates throughout the world have long been considered prey for other species. This assumption has some basis in the large numbers of monkeys and apes that are captured or killed annually by men who operate within a range of motives that include monetary profit and scientific research. But a continuing paucity of actual observations in natural habitats, despite the increasing attention given to wild primates in recent years, leaves some doubt as to the degree to which primates serve as prey to predators other than man.

Chimpanzees are reputedly preyed upon by leopards and other large carnivores, but no actual incidents have been reported from field studies in the Budongo Forest (Reynolds and Reynolds, 1965; Sugiyama, 1968), the Kasakati Basin (Izawa and Itani, 1966; Suzuki, 1969), and Gombe National Park (Lawick-Goodall, 1968b), or in Central Africa (Kortlandt, 1968b, manuscript) and West Africa (Jones and Pi Sabater, 1971). At Gombe, and perhaps elsewhere as well, predation seems to be of minor importance in the annual reduction of the chimpanzee population. Diseases (e.g. polio) and accidents (e.g. falling from trees) create greater inroads: these factors accounted for all 6 chimpanzees lost at Gombe during 1968–69. Similarly, mountain gorillas may occasionally be killed by leopards, but again no direct evidence to this effect has been reported (Bingham, 1932; Kawai and Mizuhara, 1959; Schaller, 1963, 1965a–b). More recent field studies of highland gorillas by D. Fossey (1970, 1971) in East Africa may produce additional information on predation. Although observations of predation upon orangutans are also lacking, field surveys indicate that man constitutes the most serious danger to these apes (Carpenter, 1938; Harrison, 1955; Okano, 1965; Schaller, 1961; Yoshiba, 1964). A more extensive field study recently completed by J. MacKinnon will supply additional information. A comparable situation apparently prevails with gibbons in that reputed predators such as leopards, pythons, and eagles have not been observed capturing these small, highly agile apes (Carpenter, 1940; Ellefson, 1967, 1968).

Several field studies in Africa with drills (Gartlan, 1970), hamadryas baboons (Kummer, 1968), yellow and olive baboons (Altmann and Altmann, 1970; Devore and Hall, 1965; DeVore and Washburn, 1963; Maxim and Buettner-Janusch, 1963; Ransom and Ransom, in press; Rowell, 1966), and chacma baboons (Bolwig, 1959; Hall, 1961, 1962a–b, 1963) show that the habitats occupied by these primates are shared by numerous predatory species, including the lion, leopard, cheetah, serval, hyena, jackal and hunting dog, and a variety of raptorial birds. Nevertheless, few direct observations of successful predation upon baboons (other than by chimpanzees) have been reported from these regions.[12] In 1469 hours of observing yellow baboons in Kenya, Altmann and Altmann (1970) recorded 5 cases of predation upon these primates. Again, numerous field studies of both Old and New World monkeys—a partial list including the howler, spider, woolly, and capuchin monkeys of the Americas,[13] the macaques, langurs, and proboscis monkeys in Asia and India,[14] and the mangabey, patas, guenon, redtail, vervet, blue and colobus monkeys of Africa[15]—suggest that nonhuman predation upon these species is a rarely occurring and even more rarely observed phenomenon. Lastly, Madagascar prosimians are perhaps preyed upon by raptorial birds (Jolly, 1966; Petter, 1962, 1965), but this is mainly conjecture.

Negative evidence is, of course, not a reliable indicator of the frequency of successful predation, yet it seems likely that the amount of

12. During a month's stay in Ngorongoro Crater, Tanzania, I twice saw lioness prides isolate a few baboons in lone trees. After about an hour's futile effort of climbing into the lower branches, the big cats gave up and departed, and the baboons later rejoined their troops. Similar incidents may be common, but predator failures do not affect prey populations.

13. Altmann, 1959; Bernstein, 1964; Carpenter, 1934, 1935, 1965; Durham, in prep.; Oppenheimer, 1968.

14. Altmann, 1962, 1965; Bernstein, 1967, 1968; Furuya, 1961–62, 1962; Imanishi, 1957; Jay, 1963, 1965; Kern, 1964, 1965; Koford, 1963; Neville, 1968; Poirier, 1970; Ripley, 1965; Simonds, 1965; Sugiyama, 1967; Yoshiba, 1967, 1968.

15. Aldrich-Blake, in prep.; Bourliere, Hunkeler and Bertrand, 1970; Buxton, 1952; Chalmers, 1967, 1968a–b; Clutton-Brock, in prep.; Gartlan, 1968; Hall and Gartlan, 1965; Struhsaker, 1967a–b.

effective predation upon apes and monkeys throughout the world has been somewhat overrated. To complicate the issue additionally, there is always a possibility that certain uncontrolled variables, such as the disruption introduced by an observer who has not conditioned potential predators, appreciably affects the current picture of predation upon primates. Even the decimation of natural populations by man continues at an unknown rate in most regions. At present, humans probably do not pose a serious threat to apes inhabiting Uganda and Tanzania, where many protective measures have recently been taken, but gorillas and chimpanzees are still hunted, eaten, and collected in parts of Central and West Africa (Morris and Morris, 1966).

Numerous tales of men, women and particularly children being terrorized or captured by fearsome manlike creatures have for more than a century emerged from the tropical forests of Africa and Asia (see, for example, Hartmann, 1886, or Morris and Morris, 1966). Gorillas, chimpanzees, and orangutans would, accordingly, qualify as vicious and dedicated predators at least insofar as human prey were concerned. The veracity of such stories is often questionable. Only one relatively credible incident of a human being mauled by chimpanzees has been recorded at Gombe, when a man with a badly scarred face appeared at the station and claimed that an adult male chimpanzee injured him as a child when he ran to save his younger brother from being abducted. Although reports of monkeys attacking humans are rare, there are several cases on record of baboons killing and eating both wild game and domestic stock (Cullen, 1969; Dart, 1963; Hall, 1966). At the turn of the century E. Marais (1969) noted that, although he himself never saw baboons naturally eat the flesh of mammals, baboons near the Cape regularly killed domestic sheep and obtained curdled milk from their stomachs. Other reports expand the prey list to a variety of wild mammals. An exceptionally detailed report by H.B. Potter, game conservator in Zululand for 20 years, describes baboons killing a variety of wild prey for food (Oakley, 1962).

PRIMATES AS MEAT-EATERS AND PREDATORS

That the general primate diet includes eggs and such small animals as nestling birds, rodents, reptiles, and insects has long been suspected and has recently been often observed in natural conditions (Washburn and Hamburg, 1968). Some examples among the smaller monkeys include vervets eating rodents (Gartlan and Brain, 1968), woolly monkeys eating tree rats (Durham, per. comm.), and capuchin monkeys eating newborn rodents (Oppenheimer, per. comm.). In Gombe National Park chimpanzees have been observed raiding bird nests and catching a variety of insects, and baboons have been seen eating fish, eggs, lizards, and insects. During the early 1960s, field evidence of primates capturing and eating the young of larger mammals began to accumulate from several sources in East Africa. By the close of the decade a list of such episodes, notably among baboons (Altmann and Altmann, 1970; Bartlett and Bartlett, 1961; DeVore and Washburn, 1963; Lawick-Goodall, 1967a) and chimpanzees (Goodall, 1963b; Kawabe, 1966; Lawick-Goodall, 1968b; Sugiyama, 1968), served to establish firmly the activities of capturing and consuming various mammals as a natural form of nonhuman primate behavior. Predatory episodes were photographed,[16] and thus observations were confirmed.

In 1966, Hall summarized 21 cases of meat-eating by baboons in various regions of Africa; 3 gazelles, 13 hares, 4 vervet monkeys, and 1 bush baby were captured. More specifically, one 12-month field study of olive baboons living in Nairobi Park, Kenya, yielded information on 5 cases of "opportunistic" predation (DeVore and Washburn, 1963). Twice the prey were young hares, twice very young Thomson's gazelles, and once a juvenile vervet monkey. Several fledgling birds were also collected and eaten. All these incidents involved adult male baboons, none of whom voluntarily shared meat with other interested individuals. All were "opportunistic" in the sense that no systematic searching or stalking activities preceded the captures. More recent observations on yellow baboons by Altmann and Altmann (1970), who claim that all components of predation except pursuit are present among baboons, also support this premise. Together with T. Struhsaker, these researchers recorded a total of 7 chases that failed, on 3 Cape hare and 4 vervet monkeys, and 23 kills that included 9 Cape hare, 2 bush babies, 11 vervets, and 1 neonatal Grant's gazelle. And in a 1-year study of hamadryas baboons in Ethiopia, Kummer (1968) once observed a female carrying a freshly killed young dik-dik antelope.

Several elements in the Nairobi Park cases (DeVore and Washburn, 1963) are of comparative interest in regard to predation by Gombe chimpanzees. The first concerns the case of a male baboon

16. Photographs of meat-eating by primates have appeared in Goodall (1963) or Lawick-Goodall (1967a, 1968b), Dart (1963), DeVore and Washburn (1963), DeVore and Hall (1965), Eimerl and DeVore (1965), Leakey (1969), and Altmann and Altmann (1970).

who had just captured a hare and, when harassed by 2 eagles, dropped pieces of the ribcage and foreleg which were then ignored by 2 other baboons who had been showing interest in the carcass. Moreover, baboons were not seen scavenging from other kills during the year. Among chimpanzees, lack of response to proferred meat has been recorded by Kortlandt (1966, 1967) in Central Africa, and at Gombe carcasses and fresh meat of individuals they had not killed is ignored. Baboons that died naturally at Gombe sometimes aroused curiosity among the chimpanzees but were not eaten (Lawick-Goodall, 1968b). Once a kill was made, however, discarded pieces were definitely not ignored by other chimpanzees at Gombe. Another Nairobi Park case, the killing of a young gazelle, holds special interest in that (a) the male baboon caught the infant, raised it overhead, slammed it to the ground, and began to eat by tearing into the stomach; and (b) the viscera were consumed first, then the flesh, and finally the skull was cracked open, and the brain scooped out by the fingers. These patterns of killing and eating, and even the technique of removing the brain, were several times observed among the Gombe chimpanzees. The third, fourth, and last correlations are simply that both baboons and chimpanzees apparently focus on the young of other mammalian species, that both consider other primate species as potential prey, and that mainly adult males capture prey.

Although Gombe baboons killed and ate a young bushbuck in 1964, both troops under regular observation from October 1967 to April 1969 were not seen to chase or eat mammals. Members of both troops were, however, observed eating dried fish on the beaches, and perhaps some baboons even collected aquatic food when walking in streams and small pools. Although baboons regularly passed with disinterest near live domestic chickens and ducks, adult male baboons twice stole and ate freshly plucked whole chickens from the camp kitchen.

It is important to note that such extensive data as have been accumulating in Gombe National Park have not been obtained from other regions. Examination of chimpanzee dung in the Kasakati Basin area shows, however, that meat-eating occurs there as well (Suzuki, 1969), and one case of predation upon a redtail monkey was reported in detail by Kawabe (1966). The activities observed during that episode correspond closely with the predatory behavior seen at Gombe. Furthermore, A. Suzuki has reportedly observed chimpanzees eating a blue monkey and a black-and-white colobus monkey in Budongo Forest (Sugiyama, 1968). Active predation has, therefore, been documented in 3 separate populations living in 2 ecosystems—the savanna woodland in Tanzania and the tropical forest in Uganda.

Wild chimpanzees are certainly omnivores in that a great variety of foods—including leaves and shoots, buds, bark and resin, fruits and blossoms, berries, seeds and nuts, reed pith, honey, insects, eggs, meat, and minerals (Lawick-Goodall, 1968b) —are consumed in a regular or cyclic fashion each year. However, chimpanzees have also been classified as frugivores and vegetarians. All these terms are descriptive of dietary habits, but each applies within different limits, or reference frames. Thus, chimpanzees may be considered frugivorous inasmuch as fruit probably composes the bulk of their intake during at least some months of each year, and they are also vegetarian because plant materials are regularly consumed in greater variety than any other class of foods. To this list must now be added the fact that chimpanzees, in at least some regions of Africa, are predacious, for live mammals are caught in a deliberate and systematic manner. In comparison to the "opportunistic" meat-eating reported from savanna baboon studies, reports on predatory behavior from Gombe and Kasakati Basin are qualitatively unique in that a stage of pursuit—chasing or stalking—is clearly present.

Whether the activities involved in acquiring insects, nestling birds, newborn rodents and the like should be included within the scope of active predation is a matter of definition. Viewed as an active feeding mechanism rather than a referent to the type of food being consumed (Bates, 1958) , a distinction can be made between predatory and collecting behavior in which the acquisition of most small animals falls into the latter category. The above distinction is adopted in this report because a chimpanzee that raids a weaver bird nest or opens an ant hill exhibits a form of collecting activity that is similar to the foraging techniques used in gathering many plant foods, and dissimilar from the sequential activities of pursuit, capture, and consumption exhibited during predation upon mammals. The meat collection usually observed among baboons and other monkeys may nonetheless be important as a potential base for the development of actual prey pursuit.

THE GOMBE PREDATION DATA

A total of 95 cases of *successful* predation was recorded at the Gombe Stream Research Centre between 1960 and 1970, all within that portion of the park used heavily by the study community. A number of additional cases was observed among unhabituated chimpanzees some distance from the main study area, but these are not tabulated because no members of the study community were present. Not all of these cases were observed in the form of meat-eating episodes: successful predation was *observed* in 46 instances, chimpanzee *feces* yielded the remains of 38 prey individuals, and chimpanzees brought the *remains* of 11 prey to the feeding area. In 56 of the 95 kills the prey species was identified (Table 1). Definite evidence thus exists that at least 6 mammalian species can be prey for chimpanzees, and these include nearly all of the species potentially available in the park.[17] With the exception of red colobus and redtail monkeys, only

TABLE 1

Prey Identified Between
June 1960 and August 1970

Prey	Tally	% of Tot.	% by Kind
Olive baboon	21	38	65
Red Colobus monkey	14	25	
Blue monkey	1	1	
Redtail monkey	1	1	
Bush pig	10	19	35
Bushbuck	9	16	
Total	56	100	100

the young of these species was captured. These figures also indicate that other primates are the most frequent prey.

In addition to the above total of 95 cases, 37 *unsuccessful* episodes were recorded during the same years. Because observation at Gombe has not always been continuous, because staffing fluctuates yearly, and because unsuccessful predation is less likely to be noticed consistently, these figures may well be distorted in that they probably show a lower-than-actual estimate of the number of episodes which occurred in 10 years, and they provide a poor estimate of the predatory success rate (95/132=73%) of chimpanzees.

During the last 30 months of research, a period between March 1968 and August 1970 when observation around the feeding area was highly consistent, 44 predatory episodes—21 kills and 23 attempts—

17. Vervet monkeys are absent from prey lists probably because (a) they are more rare in the park than are other primates and (b) they inhabit primarily beach zones that are not frequented by chimpanzees.

were recorded within lower Kakombe Valley (Table 2). The total number of kills is somewhat small for correlation with the seasons, but a tendency for slightly more kills during the rainy months is apparent (fig. 1).

Breaking this 30-month period into sections of 12 months and 18 months respectively, there were 12 kills in the first period and 9 in the second. This decrease of 50% (from 12K/12M to 9K/18M) was

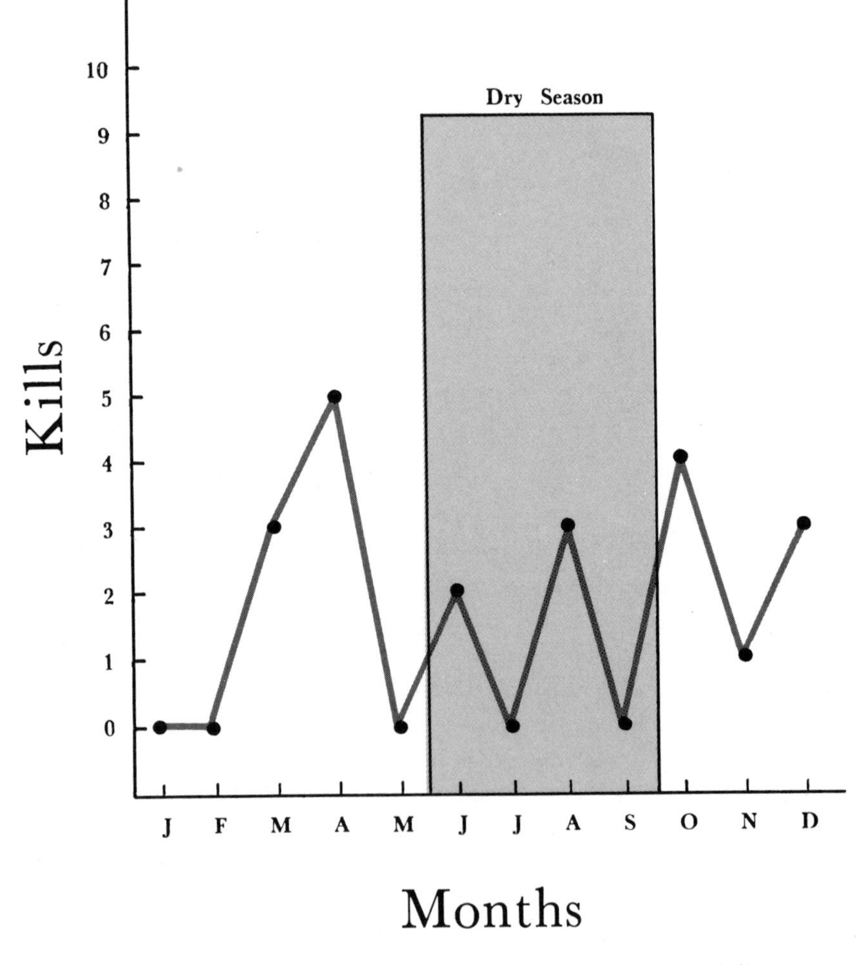

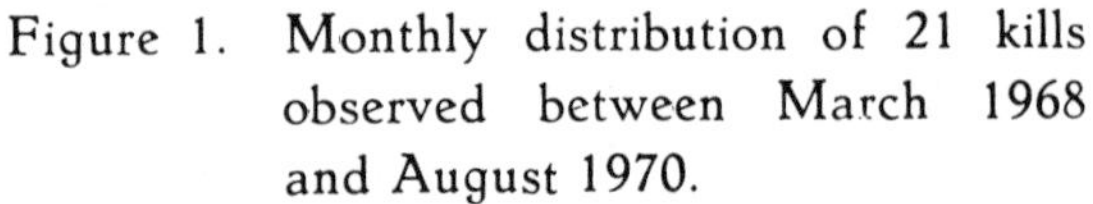
Figure 1. Monthly distribution of 21 kills observed between March 1968 and August 1970.

TABLE 2

Prey Identified Between
March 1968 and August 1970

Prey	Kills	Attempts	Tally
Olive baboon	12	22	34
Red Colobus	4	—	4
Blue monkey	—	—	—
Redtail monkey	—	1	1
Bush pig	3	—	3
Bushbuck	2	—	2
Totals	21	23	44

perhaps connected with the feeding reduction initiated in June 1968. However, the apparent decrease in the rate of predation is probably better explained by less consistent observation resulting from the lower chimpanzee attendance rate that accompanied the reduction of banana distribution. The difference between these periods is further marked by the fact that baboons constituted 83% of the prey taken in 1968–69, when both species were regularly together at the feeding area nearly every day, but only 22% of those prey captured in 1969–70. A comparison of kill rates in these periods shows that banana feeding probably affected the type of prey captured more than the yearly rate of predation.

Focusing more closely upon the period from March 1968 to March 1969, Gombe researchers recorded in detail 30 cases of predatory behavior.[18] A total of 47 observation hours was expended on predatory activities. Chimpanzees brought the remains of 3 carcasses to the feeding area, and unfamiliar chimpanzees were once seen eating meat in a distant valley; but these 4 cases are excluded from subsequent discussion because behavioral data were lacking. All of the 30 observed episodes occurred in the lower portion of Kakombe Valley, most within 150 yards of the feeding area (Appendix A, Map V). These episodes involving the chimpanzee study community may constitute only part of the actual 1968–69 complement, for some cases may not have been recognized and others are likely to have occurred in other valleys.

That baboons were preyed upon most frequently and, conversely, that a red colobus and a bushbuck were captured only once each

18. As a participant in general research from March 10, 1968, to the end of February 1969, I was present at most incidents in the capacity of recorder, photographer, or standby observer.

during 1968–69 may be the result of several factors: (a) greater concentration of researchers in the vicinity of the base camp than elsewhere, so observational efficiency was correspondingly greater in Lower Kakombe Valley; (b) concentrations of baboons and chimpanzees were, especially at times affected by feeding, higher than elsewhere in the normal ranges of these primates and, conversely, these concentrations (and the excitement accompanying banana destribution) may have reduced the incidence of other potential prey species in the area during daylight hours; and (c) the emphasis upon baboon prey may well have been associated with a temporary preferential phase for baboon meat, or simply the peak of a dietary cycle similar to those which occur yearly with other kinds of foods such as insects and fruits.

Only 12 of the 30 episodes, or 40%, led to *kills* followed by consumption of the prey. Considering baboon prey alone, 10 out of 28 episodes, or 36%, were successful. The remaining 18 episodes were *attempts* because of failure to capture the selected prey individual in spite of the clear predatory behavior of one or more chimpanzees. The average duration of kills was 226 minutes, with a range from 105 to 540 minutes; the average for attempts was 12 minutes, also with a wide range from 1 to 71 minutes. Episodes may be conveniently divided into basic activity stages: (1) the pursuit, (2) the capture, and (3) the consumption. Average durations for these periods were 9 minutes, 2 minutes and 215 minutes respectively, although the length of each was highly variable. The first stage may in turn be separated into several behaviorally distinct modes of pursuit: (1) seizure, (2) chase, and (3) stalk. These modes occurred in a ratio of 7:11:4 respectively, with 8 cases remaining undetermined. Taking each mode independently the success rates were 43% in seizures, 55% in chases, and 00% in stalking.

During the 1968–69 study period, only adult male chimpanzees were observed to initiate predation and to pursue and capture prey (see Appendix B, Table II). Females and subadults sometimes followed such males toward the prey at distances of a few yards but did not join directly in the activities until consumption was underway. Data from other years show that this class distinction in behavior is not absolute, for females and adolescents do on occasion pursue and even capture prey, especially if adult males are not present. Though the 1968–69 period may in this respect have been somewhat exceptional, it is nevertheless probable that prey procurement is, unlike the sharing and eating of meat, primarily an adult male activity.

Estimates concerning the potential impact of chimpanzees as predators are worth considering here. Basing calculations on the 95 kills recorded in 10 years, an average emerges of nearly 10 prey taken each year. Assuming that the rate of success during the entire decade is comparable to the 40% success rate recorded in 1968–69, then the total number of postulated episodes, including attempts, would be close to 250, or about 25 per year. Allowing for a total park population only three times the size of the study community, about 75 predatory episodes might reasonably be expected to occur per year within an area of 25–30 square miles. Moreover, there is no assurance that all episodes were observed and recorded; this figure may represent only a small part of the true total.

Turning more specifically to successful predation, a second estimate can be obtained. If about 10 prey individuals are removed from the study area by about 50 chimpanzees, then 150 chimpanzees may well be successful in killing 30 or more prey of mixed species in the entire park within an equal time period. Alternatively, this would mean about 300 kills and about 450 attempts per decade within the park.

Thirdly, a behavioral estimate of the extent of chimpanzee predation can be calculated from data on individual participation rates. Within the study community of 50 chimpanzees, individuals were involved in predation nearly 200 times during the 1968–69 period. If this figure is extrapolated to incorporate the probable population (ca. 150) of the entire park as well as the postulated number of episodes (ca. 75) that might occur in a year, then the known figure of 200 expands to 1500 separate instances of participation per year within the entire park.

Shifting finally to the viewpoint of the prey captured in 1968–69, and using the 2 baboon troops as examples, the fact emerges that 7% of Camp Troop (membership 75) and 9% of Beach Troop (membership 55) were eliminated in that year by chimpanzee predation alone. In the specific case of Beach Troop, the 5 captures observed accounted for 29% of the lowest age group—i.e. baboons between birth and 2 years of age.

Even if these various calculations are slightly higher than actual the possible effects of chimpanzee predation upon the population dynamics of the park appear unexpectedly high for a species hitherto considered "nonpredatory." It seems likely that prey killed by Gombe chimpanzees add a significant number to the prey killed by other, more specialized carnivores inhabiting the same area.

General comparison of the 1968–69 data on chimpanzees to information accumulating from field studies on other predators is additionally interesting. Together with a few comparative reports (Wright, 1960; Kruuk and Turner, 1967), several systematic ecological studies of East African carnivores—such as the cheetah (Schaller, 1968), the lion (Schaller, 1969a-b; Makacha and Schaller, 1969), the spotted hyena (Kruuk, 1966, 1968), and the wild hunting dog (Kuhme, 1965; Estes and Goddard, 1967)—provide pertinent data on numerous aspects of predation. Schaller discovered that the predatory success rate of cheetah was about 50%, that 337 prey might be captured per year by an adult who consumed about 9 pounds of meat per day. Schaller also calculated that the success rate of lions was 52% with 2 or more participating, but only 29% with solitary lions, though added complications arise in that rates may drop in special environmental conditions. Additionally, Wright estimates that an average pride of 3 adult and 3 subadult lions captures about 219 prey per year, with an adult consuming about 40–45 pounds per day. Among hyenas, Kruuk observed a predatory success rate for packs of about 70%, and a much lower rate for solitary hyenas. In the case of wild hunting dogs packs, Estes and Goddard reported an exceptionally high success rate over 85%, with about 6 pounds of meat consumed per day by an adult. However, the success of these dogs hunting small prey, such as gazelles, is probably considerably lower, according to Lawick-Goodall (1970), who calculated a success rate of 43% from 91 observed episodes. Although actual observation of predatory episodes is rare among timber wolves, field studies of these carnivores (Mech, 1966; Pimlott, 1967; Pimlott, Shannon and Kolensky, 1969) indicate that their success rate may at times be as low as 9% with large prey such as moose.

Allowing for observational discrepancies and methodological differences in various field studies, the success rate of Gombe chimpanzees remains in reasonable accord with the rates of several carnivores (Table 3). The actual amount of killing per year and the volume of meat consumption per individual are, of course, much lower among chimpanzees. Even if these chimpanzees are considered predators on the basis of success rate, they do not qualify as carnivores because fresh meat composes a negligible portion of their total caloric intake. Meat comprised less than 1% of the yearly diet among the baboons observed by Washburn and DeVore (1961a), and it is unlikely that any given chimpanzee, even an adult male, consumes very much more.

In a broader perspective, the Gombe chimpanzees may qualify as omnivorous forager-predators not entirely dissimilar, at least in terms

TABLE 3

Estimated Success Rates of
Gombe Chimpanzees and Several Plains Predators

Predator	Solitary Rate	Group Rate
Chimpanzee	?	40%
Cheetah	50%	?
Lion	29%	52%
Hyena	70%	70%
Wild Dog	?	43%

of food-getting behavior, from the current view of early man as living in small communities of hunter-gatherers. Indeed, basic similarities between this chimpanzee study community and certain prehistoric human communities may well include subsistence on similar classes of foods obtained in basically comparable ways. Specific data on hunting activities among marginal human groups (e.g. success rates, meat intake) are so limited that it is difficult to compare humans, chimpanzees, and carnivores. If Woodburn's (1968) and Marshall's (1965) estimates that about 20% of the bulk intake among the Hadza of Tanzania and !Kung Bushmen of the Kalahari is provided by the male hunters are generally true of hunter-gatherer groups, then the Gombe chimpanzees do not rank with human hunters.

2

SELECTED PREDATORY EPISODES

All field studies of primate behavior have qualitative and, therefore, selective undertones. These are manifested more than usually in the study of chimpanzees, who often appear to the human observer exceptionally complex and subtle in their behavior.[1] An exclusively analytical presentation would thus be incomplete without a description of the conditions and activities comprising predatory episodes as seen by the observers. The purpose of this chapter is to provide perspective, particularly for those lacking direct familiarity with the behavior of wild chimpanzees, within which to assess the original data and subsequent analysis.

Six predatory episodes, including 3 attempts and 3 kills, have been selected from the total complement of 30 seen in 1968–69, and these are presented in the descriptive, narrative style of the original field journal, which has only been slightly condensed and reworded to enhance readability. Predation upon mammals other than baboons has been excluded, but excerpts from such episodes are provided in the following chapter. In addition to illustrating diverse conditions of predation, each of the 6 cases includes important components of predatory behavior that will be discussed in the following chapter. The great diversity of conditions and activities covered by these few episodes may in fact be the most significant aspect of chimpanzee predatory behavior.

1. For example, the chimpanzee excitement accompanying predation usually emanates to human observers. Full understanding of such "semiparticipant" responses is still lacking, and observational acuity may either increase or decrease as a result.

KILL: MARCH 19, 1968

Observing conditions during this episode were excellent in the initial stages because it took place directly in the feeding area, and 4 researchers (Gale, Sorem, T. Ransom, Teleki) collaborated on the final notes. An excellent example of seizure, this episode shows how a single chimpanzee can successfully capture a young baboon in the midst of the baboon troop, which retaliates by mobbing the predator.

Since boxes have been filled with bananas the previous evening, banana distribution begins at 7:11 A.M. as the first chimpanzees arrive. Only a few boxes are open by the time Camp Troop baboons appear at 7:15. Hugh (adult ♂), Mike (adult ♂), and a few others continue eating bananas. As the boxes cannot be closed, there is no alternative but to let the chimpanzees finish even as baboons of all ages stream into the feeding area. By 7:40 most of the troop (ca. 60) is here, so the remaining boxes are kept shut in the hope that at least a part of the troop will leave shortly.

During the next half hour, while Mike eats nearly a whole box full of bananas (ca. 25), chimpanzees continue arriving from different directions until, by 8:15, 18 are in camp: 9 adult males,[2] 2 adult females, 2 mothers with infants, 1 adolescent, and 2 juveniles. Light rain has been falling steadily since early morning, so many of the waiting chimpanzees are dispersed in trees. In one south slope tree sit 3 adult males—Mike, Charlie, and Hugh; sitting on a box directly below the males, mother baboon Arwen holds her infant Amber closely ventral. Only a few baboons have drifted away from camp; most of the troop remains concentrated in the south and southwest sectors of the feeding area, the same sectors occupied by most of the chimpanzees.

Several chimpanzees, particularly the adult males, become increasingly agitated (or frustrated) as the minutes pass: knowing the boxes are full of bananas, all are expectant. In spite of this growing tension, feeding is delayed a bit longer in order to avoid the chaos that would result from having more than 60 individuals of both species present.

At approximately 8:20, tension suddenly erupts in action as Leakey (adult ♂) attacks Figan (adult ♂). Possibly taking advantage of the ensuing uproar and general distraction, Mike chooses this moment to quickly leave Charlie and Hugh on their branch and, with a flick of his right hand as he steps down from the tree trunk behind Arwen

2. MK, LK, HH, HM, RX, WZ, CH, EV, FG

baboon, grabs infant Amber. Breaking into a bipedal run slightly up and then across the south slope, Mike flails Amber overhead, twice striking her against the grassy ground.

Mike's actions instantly elicit a burst of excited *screaming* from chimpanzees and *barking* from baboons which drop from trees all around camp.[3] In a matter of seconds Mike covers about 15 yards while male chimpanzees and baboons converge from many directions. *Screeching* repeatedly, Arwen baboon darts about in the melee. Perhaps so that he can ward off mobbing baboons with both hands, Mike transfers Amber to his mouth while continuing his bipedal, arms-flailing advance. Still very much alive, Amber kicks her arms and legs as Mike runs, teeth clamped across her spine; she may be vocalizing as well, but the general din blankets any possible sounds from her.

Mike's speed slows considerably as a number of milling chimpanzees and baboons (ca. 12 by now) impede his movement across the open slope. No direct attacks on Mike are noted, but action is so fast that determining who attacks or aids whom is impossible: baboons lunge, canine-threaten, and *bark* at chimpanzees, and possibly at each other, while chimpanzees swagger bipedally, wheel, and slap out repeatedly at baboons and one another. Seconds later Mike halts in the center of the melee, standing bipedally just a few paces downslope from the veranda. Baboons and chimpanzees immediately close in until, after a second's pause, Mike again moves bipedally across the slope flailing both arms at those nearest. At least 3 male baboons harry him closely as the milling throng approaches a patch of taller grass (ca. 5 feet high) near the southwest edge of the feeding area.

Suddenly one male baboon leaps onto Mike's back and, riding there by gripping hair with hands and feet, repeatedly rakes his open mouth across Mike's shoulders.[4] Stopping abruptly, staggering, Mike takes Amber baboon into his left hand; while still bipedal, Mike rapidly rotates his body from side to side and violently waves both extended arms at shoulder level in an effort to dislodge his assailant. Mike smashes the back of his left hand against the trunk of a nearby sapling during one rotation, the impact breaks his grip, and Amber presumably flies into the tall grass.

3. Although no standard terminology exists for baboon and chimpanzee expressions and vocalizations, the descriptive field terms of Hall and DeVore (1965) for baboons and of Lawick-Goodall (1968b) for chimpanzees are used where possible in this text. All terms describing sounds are emphasized by italics.

4. Later examination showed Mike's back was not even slightly wounded. Though the aggressor's identity remained uncertain, researchers fully familiar with the troop suspected a male known to be lacking canines.

Just after Amber disappears, the noise and activity begin to subside. The adult baboon drops from Mike's back. Other baboons and chimpanzees stop, appear momentarily confused as they glance about, then resume milling about but with much less interest in one another. Several enter the taller grass, crisscrossing the area while peering and poking about in the grass. A dozen individuals are soon involved in the search, some baboons and chimpanzees brushing shoulders. Tension remains relatively high, however, and aggression erupts occasionally. Among those searching, Mike soon reappears without infant Amber. Hugh, Leakey, and others reappear shortly. As several baboons emerge from the now trampled grass, one male suddenly lunges and grabs Rix (adult ♂) from behind, *barking* loudly; Rix spins around to slap the attacker, then swaggers bipedally toward the baboon with hair erect and arms akimbo; the baboon lunges at Rix once more, yawning and eyelidding, before he walks away. After this final event, chimpanzees and baboons slowly disperse from the site.

By 8:22 all is relatively calm again, though there appears to be a residue of tension among some individuals. Hugh chases Figan briefly. Mike displays at someone, then climbs back into the tree from which he started. Some chimpanzees and baboons continue pacing about the grassy slope, and one or another occasionally traverses the patch of grass where Amber disappeared. The mother baboon, Arwen, is the last to leave the site.

At about 9:00 the rain stops and a small cluster of chimpanzees is noticed some 200 yards south of the attempt site. Minutes later chimpanzee *screams* are heard there, and 6 male chimpanzees, Mike among them, quickly leave the feeding area. Humphrey (adult ♂), who was not a central participant in the earlier capture excitement, is identified in a tree near the base of Sleeping Buffalo Ridge; sitting alone on a branch, he holds a limp, presumably dead baboon in his hands. Reconstructing events later, observers surmised that Humphrey was just on the other side of the grass patch where Mike dropped Amber baboon, and that he probably snatched up the infant and hurried south before the others noticed. Mike, Charlie, Leakey, Hugh, and others arrive at Humphrey's tree at 9:05, and the meat-eating commences. Due to poor visibility the observers are unable to determine the exact time taken for consumption of the carcass, but 10:26 A.M. marks the last occasion that Mike is seen chewing meat. A group grooming session follows, lasting well over an hour. Nearly all the chimpanzees return to camp by 1:32 P.M., at which time banana feeding is resumed.

ATTEMPTS (2): APRIL 27, 1968

Two researchers (Gale, Teleki) observed these episodes from a few yards distance and collaborated on the final notes. Observing conditions were excellent, except at brief intervals when a building obstructed the view. The 2 cases in one day show that consecutive attempts do occur sometimes. The number of chimpanzee participants in the second episode is of some importance because at no other time during the year were so many seen stalking collectively. Although the first case cannot be classified accurately, the second qualifies highly as a stalking attempt.

The day begins at 7:00 A.M. as researchers open boxes for the first chimpanzees. Beach Troop baboons arrive at 7:38, but banana feeding continues because most of the chimpanzees already occupy boxes. By 9:00, after several cases of interspecific aggression are recorded, no new boxes are opened. Baboons soon start drifting out of the feeding area, but more chimpanzees arrive and feeding resumes at 10:30. About a half hour later most of the troop returns, and banana distribution again stops. By this time all the chimpanzees present have been fed: 12 adult males,[5] 3 adult females, 5 mothers with infants, 3 adolescents, and 1 juvenile. Many baboons again leave by about 11:15, though some stay on the open grass slopes. The situation is quite calm by 12:30 P.M. Baboons wander about slowly and most chimpanzees are in trees, grooming or in day nests, because a light rain has been falling since early morning.

Figan (adult ♂), who has been sitting idly in a tree, suddenly drops to the ground at 12:32 P.M. and hurries silently across the open slope toward a small baboon cluster—an adult male, a female, and a young juvenile. Almost simultaneously, Rix (adult ♂) and Worzle (adult ♂) descend from other trees to join Figan, and all stop about 5 yards away and watch the same baboons. Figan, standing slightly ahead of the others, slowly begins to approach the juvenile, who *gecks* (*yaks*) loudly; the male baboon immediately joins the juvenile, and both stand their ground and face the watching chimpanzees. Figan stops again about 3 yards distant. At this moment Hugh (adult ♂), Charlie (adult ♂) and Mike (adult ♂) quickly cross the slope toward the baboons with hair on end; the juvenile baboon *screeches* and *gecks* loudly, the male lunges and eyelids; Charlie stands, arm-waves, and swaggers toward the cluster. The baboons immediately retreat to the

5. MK, LK, HG, HH, HM, RX, WZ, FB, PP, CH, EV, FG

edge of camp. Only a minute after the incident has begun the chimpanzees start dispersing. The last of the baboons leave during the next few minutes.

Several chimpanzees again show interest in bananas, so feeding resumes at 12:52. At least half of Beach Troop (ca. 30) returns rapidly when the chimpanzees vocalize excitedly. There is only 1 recorded case of interspecific aggression, but so many individuals are present that operations again cease at 1:07. Having been fed recently, most of the chimpanzees calm down by 2:00, and most climb into nearby trees to sit out the mild rain. Very few baboons remain in sight, but some stay just west and southwest.

Hugh (adult ♂), Hugo (adult ♂), and Charlie (adult ♂) have for some time been sitting quietly on the west slope of the feeding area. At 2:18 Hugo rises and starts to walk along a path near a building; Hugh and Charlie follow part way, then sit down and watch Hugo, who begins pacing back and forth along the path. Mike (adult ♂) rapidly and silently descends from a nearby acacia tree, stops some yards from the building, and also watches Hugo. Humphrey (adult ♂) approaches from the west, stops to copulate with Fifi, and then sits a dozen yards away, also watching.

Shortly several of the watching males—Hugh, Charlie, and Mike—quickly and silently join Hugo in pacing around the building. They pass from sight at 2:20, and an infant baboon *chirps* (*chirplike clicking*) somewhere. A male and female baboon, the latter carrying her 1-week-old black infant, Thor, then appear in a tree overhanging the building.

Humphrey suddenly runs downslope and joins Mike, who is alone near the north wall. At the same time Hugo, Hugh, and Charlie continue pacing along the south wall. Only Mike and Humphrey stare fixedly up and over the roof at the baboons; the others glance upward constantly while moving about. Satan (adolescent ♂) slowly approaches within 20 yards of the other males, climbs a tree, and watches intently from there. No other chimpanzees appear at the predation site. By now the 5 adult males have effectively surrounded the building, and each member of the stalking cluster moves with such coordination that each wall is covered by someone at any given moment. All maintain complete silence.

By 2:26 all but one chimpanzee have stopped moving but continue to stare up at the baboons: Hugo alone continues to pace around the building. The male baboon sits still on the branch, staring at one and then another chimpanzee below; only inches away, the female repeatedly glances at the chimpanzees below and makes rapid, agitated movements with her hands and body toward the tree trunk. Though the male baboon appears unperturbed, the female attempts to leave. She holds infant Thor, who is now silent, tightly ventral. Moments later, the male quickly stands and jumps down onto the roof, where he paces back and forth along the peak, hair erect, and tries to keep all the chimpanzees in view. He then leaps to the ground on the north side of the building, where there are momentarily no chimpanzees. But Mike turns the corner; the baboon threatens, lunging foreward with mouth wide open, canines bared. When Mike almost at the same instant swaggers and refuses to retreat, the baboon leaps back onto the roof, then up onto the branch beside the female.

Seconds after this brief clash there is an outburst of chimpanzee and baboon vocalization just behind the building. Mike is already there, and the other 4 chimpanzees instantly run to join him. The building briefly blocks the observers' view. The male baboon and 2 male chimpanzees are then seen about 10 yards away on the path leading from the feeding area. Three more chimpanzees suddenly rush past this cluster and disappear southwest into tall grass. The remaining 2 chimpanzees break past the threatening male baboon seconds later, one on either side, and follow the others into the same grass with the baboon on their heels. There is much activity just southwest, but the grass is about 6 feet high and little can be seen.

Though observation was not complete, a reconstruction of the final moments is relatively simple. The 2 adult baboons must have made a sudden break from their tree toward the west just after the male's encounter with Mike. As the mother baboon ran ahead, the male must have turned in a rear-guard attempt to hold back the pursuing chimpanzees. This maneuver, though very brief, was apparently successful with at least the first few pursuing chimpanzees.

Many chimpanzees who were peripheral to the feeding area now start moving southwest into the high grass, drawn by the sounds and activity of pursuit. A terrific uproar ensues down there, with chimpanzee *screams, grunts, waas,* and *pant-hoots* mixing with baboon *barks* and *screeches.* Trees shake and grass undulates everywhere.

Charlie is first to return at 2:27. He climbs a tree and looks southwest toward the scene of pursuit. Hugh follows shortly. Somewhat southwest of the feeding area Mike climbs a tree and lies back along a branch. There is now silence, and no movement in the grass. Calm has settled with great rapidity. The female carrying infant Thor has escaped.

Hugh remains more tense than the other chimpanzees: minutes after returning to camp he attacks Fifi. All the others sit quietly, in trees or on the ground. By 2:33, when 2 male baboons and a different female emerge from the grass, all the chimpanzees ignore them. Even when banana distribution resumes briefly at 3:30, with about 6 baboons present, aggression is minimal and no further interspecific incidents occur.

ATTEMPT: MAY 14, 1968

This episode is also the stalking mode of pursuit, yet it differs from the April 27 case in the conditions and the number of participants. The episode illustrates Figan's self-control and persistent interest in a young baboon. Three persons (Gale, B. Ransom, Teleki) were present, and visibility from a distance of about 5 yards was excellent even though part of the action occurred in the high grass just beyond the feeding area.

Another feeding day begins as researchers open boxes for the first arrivals at 6:45 A.M. Distribution occurs smoothly, with a minimum of excitement among the chimpanzees until 6 adult male baboons from Camp Troop appear at 7:05. This number increases rapidly as the rest of the troop arrives, and competition for bananas intensifies correspondingly. In a matter of minutes at least 10 cases of interspecific aggression occur. Intense excitement approaches chaos as more chimpanzees and baboons approach steadily from many directions. Boxes are kept shut for a short period, then some are opened again at 7:25 when most individuals calm down considerably. But at least half the baboon troop (ca. 35) remains in camp, some male baboons harassing female chimpanzees for bananas, and distribution of bananas stops again at 8:11. Sporadic feeding then continues throughout the morning hours as the researchers attempt to keep competition at a low level without creating undue frustration among the chimpanzees.

By 9:35 many chimpanzees are present: 12 adult males,[6] 3 adult females, 6 mothers with infants, 2 adolescents, and 2 juveniles. Nearly the entire baboon troop (ca. 60) is also in camp by this time. However, the situation remains considerably more calm this time than it was during the early stages of feeding; all the chimpanzees have been fed at least once and most of the males appear satiated for the moment.

6. MK, GOL, LK, HG, HH, HM, RX, WZ, PP, CH, EV, FG

Only a few of the 40 boxes still contain bananas. All the chimpanzees appear relaxed, many resting in trees peripheral to the feeding area and others playing, grooming, or sleeping within the area. The baboons appear equally relaxed, though some still search diligently for discarded peels. The chimpanzees have shown no interest in them for the past hour or more.

The first indications of interest in predation appear at 9:40, when a young baboon *screeches* somewhere north in the underbrush and Mike (adult ♂), who has been sitting idly in the center of the feeding area, walks to the corner of the main building and looks toward the sounds. He soon loses interest but then repeats this action 4 minutes later, and peers toward baboon *screeches* from the same place. Several times the vocalizations stop, and each time Mike returns to the cluster of chimpanzees sitting near the building. Nearly an hour of silence follows, until another infant baboon *chirp-clicks* repeatedly at 10:35, this time somewhere just southwest of the feeding area. Mike immediately stands and looks intently downslope toward the sounds; Figan (adult ♂) rises seconds later and walks toward the baboons. Two young chimpanzees, Flint (infant ♂) and Goblin (infant ♂), join Figan and also look as they stand beside him but quickly lose interest and resume play. Two minutes later Figan walks another few yards toward the baboon sounds, then *screams* as a male baboon threatens him from the higher grass. Watching from the veranda, Mike and Goliath (adult ♂) stand and the former runs a few yards toward Figan. Both stare toward the baboons, who now move away slightly to a distance of about 6 yards. Figan sits down and Mike returns to Goliath. At 10:40 the baboon infant resumes *chirping* and Mike instantly runs downslope to Figan; when the infant quiets down seconds later, Mike again returns to Goliath. Figan sits quite still this time and stares toward the baboons.

Minutes later these sporadic indications of interest, stimulated in each case by vocalizations from the baboon infant, become more intense. At 10:44 Figan suddenly stands and walks slowly and silently toward the baboon cluster, which includes an adult male, 2-week-old black infant, Lamb, and an adult female. Held ventral by the male baboon, Lamb struggles and *chirps* repeatedly as she tries to return to her mother. When Figan stands bipedally about 5 yards away and looks intently at the trio, Mike again hurries downslope and sits nearby. Both stare fixedly at the baboons, who in turn keep glancing at the chimpanzees.

A few feet from the male chimpanzees, Figan's sister Fifi (adult

♀) plays with younger brother Flint, who *laughs* loudly as they wrestle in the grass. Their mother Flo (adult ♀) is stretched out on her back nearby, a hand shading her eyes. Soon Figan steps closer to the baboons and sits again, glancing at them occasionally. Only the female baboon appears disturbed by the chimpanzees. The male baboon's attention remains focused on infant Lamb, who continues her struggles.

At 10:48, the female baboon abruptly lunges and *barks* at Figan, who raises his hair slightly, rocks slowly back and forth on his haunches, and looks away. Mike sits quite still and continues staring.

Only 3 or 4 yards from the chimpanzees, the male baboon repeatedly lifts infant Lamb from the ground, turning her about while licking and lip-smacking at her. Infant Lamb struggles incessantly, apparently trying to get free; the female ignores Lamb and continues giving most of her attention to Figan, who ceases to rock, hunches forward suddenly and stands. He then walks slowly and silently, with very deliberate movements, in a clockwise direction around the baboon cluster, always maintaining his distance. The female baboon instantly stands, steps back a few paces, and *barks* at him; still handling the *chirping* infant, the male seems to ignore Figan completely. Mike sits quite still, watching intently from the background; Fifi continues wrestling with Flint, and Flo now appears asleep.

Figan soon stops about 2 yards from the baboons, who are now directly between the 2 male chimpanzees. During Figan's circular maneuver the female baboon has backed away slightly, leaving the male and Lamb. Her excitement now seems to increase as she threatens with her eyelids and with rapid, jerky movements of her body and arms toward Figan. Only occasionally does she glance at Mike, who continues staring fixedly. The male, still attentive to Lamb, does not even look up.

Then, at 10:50, Fifi suddenly stands, leaves Flint, and ambles straight toward the baboon cluster, seemingly unaware of the situation. The female baboon immediately stands and *barks* loudly at her. This time the male baboon looks up, takes Lamb ventral, and moves closer to the female and away from Fifi, and in the process also away from Figan and Mike. Only yards short of the baboon cluster Fifi swings up into a small tree, from which she then stares down at them; moments later the baboons move still farther away, all 3 chimpanzees staring but not moving after them. The baboons sit again perhaps 10 yards away; the male continues handling Lamb, and the female rapidly calms down.

At 10:51 Fifi swings to the ground and returns to young Flint, with whom she resumes wrestling. Figan watches the baboons a bit longer, then stands, takes a parting look at them, and walks quickly away without glancing back. Mike follows on his heels. When he reaches the veranda, Figan drops to his haunches and, head canted slightly back, glares fixedly toward his family (Flo, Fifi, Flint) for nearly a minute.

Banana distribution resumes at 11:09, and the last box is emptied about an hour later. A few cases of interspecific aggression accompany the opening of boxes, but no chimpanzee shows any sign of interest in catching a baboon.

KILL: OCTOBER 4, 1968

Capture of a baboon by a single chimpanzee crippled in one arm with polio warrants inclusion on such grounds alone, but this was also one of the few episodes of the year where detailed documentation of the actual capture was possible. There were, at varying times, at least 5 observers (Davis, Staren, T. Ransom, Sorem, Teleki) present; one person was filming and one photographing whenever possible. With the exception of a middle period when the chimpanzees were in trees, most observation and photography of this seizure was maintained from a distance of 3 yards or less. Refer to Diagram VII (Appendix C) for meat distribution sequence.

Unlike feeding days, this cloudless morning begins quietly because all the banana boxes are empty. Neither Camp nor Beach Troop baboons appear, as both are outside Kakombe Valley. A few chimpanzees are in camp by 8:30 A.M.: 3 adult males,[7] 1 adult female, and 1 adolescent. All are inactive except Mike (adult ♂) and Hugo (adult ♂), who groom mutually.

His paralysed arm swinging limply, Faben (adult ♂) departs bipedally from the feeding area at 8:55, Satan (adolescent ♂) following a few yards behind. They walk leisurely southwest on a path, soon passing a lone adult male baboon walking in the opposite direction. About 50 yards along the path both chimpanzees turn off into Palm Grove, a heavily forested section of the valley floor. Seconds after Faben enters the underbrush there are sounds of a brief, muted scuffle, then a couple of low *screeches* from a baboon. Emerging shortly from

7. MK, HG, FB

a patch of bushes, Faben stands bipedally and glances about, with his one hand holding a squirming male baboon by a leg. Suddenly he wheels and runs bipedally along a path beneath the trees, flailing the 1-year-old captive, Lane, above his head. Lane is smashed several times against nearby tree trunks, and soon becomes dazed and limp. Satan runs along some yards behind, constantly glancing over his shoulder toward the feeding area, where there are other chimpanzees. Since Lane's first outcry there have been no sounds at all.

After running about 20 yards, Faben stops in the grove and also peers intently back along the path. The next moment he is off again, half running and half leaping, and repeatedly flailing the baboon against the ground. Satan becomes increasingly nervous, then starts *squeeking* softly as he glances repeatedly back uphill toward the feeding area. Suddenly both chimpanzees emerge from the grove onto an open path just as other chimpanzees *pant-hoot* slightly uphill. Faben *pant-hoots* a response and then sits down, maintaining his grip on the baboon's ankle. Still silent, Lane baboon resumes a feeble struggle to pull away. Highly excited, Satan now joins Faben, grasps the baboon with both hands, and briefly bites the spine; being either too stupified or too intent on escaping, Lane does not defend himself.

Suddenly Hugo, Mike, and Gigi (adult ♀, estrous)[8] run down the path from camp *hooting* and *grunting* with great excitement. Satan releases Lane baboon and hastily backs away into the forest. Faben begins to *squeek* loudly and also attempts to retreat, dragging the struggling Lane behind by one foot but manages to cover only a few yards by the time Hugo arrives and sits down a few feet away; Hugo then glances rapidly back and forth between Faben and Mike, who stops with Gigi a few yards upslope and stares down at Faben. *Screaming* loudly, Faben resumes retreating across the path while trying to keep an eye on both Hugo and Mike. Just before Faben enters the forest, Mike raises his hair completely and glares down at Faben, who rushes back across the path and presents his rump to Hugo; still looking at Mike, Hugo extends one hand palm upwards and pats Faben's scrotum reassuringly.

During these activities Faben has been dragging Lane baboon about by one leg with no one else paying attention to the captive. Now, as though acting on cue, Hugo and Mike converge upon them. Faben *screams* even more loudly; Mike gently tugs on the captive a few times, and Faben releases him. The instant he has full possession of

8. Note in this and the following episode the frequency with which an estrous female participates.

Lane, Mike turns and bites down hard on the back of his neck once; Lane sags immediately, apparently dead. Hugo then joins Mike, grasps the carcass, tries to pull it away, and finally bites several times in rapid succession along the spine without tearing skin. Again both tug briefly on opposite ends, but not aggressively. After several seconds of pulling, Mike suddenly swaggers bipedally; Hugo lets go and displays toward Faben; instead of retreating, Faben *screams* even more intensely. Mike watches this activity, then sits down on the path with one hand resting on the carcass beside him. Hugo joins him, sits, and also places one hand on the baboon.

Having stayed slightly peripheral to the activity, Gigi has been *screaming* excitedly while rushing about. Satan has been watching silently and inconspicuously from the edge of the forest nearby.

Somewhat subdued at 9:00, Faben starts to walk about warily. Several times he attempts to approach the carcass but hurriedly jumps away whenever Mike or Hugo move. Suddenly Faben begins to *scream* so intensely that he develops glottal cramps and starts choking. Then the others also become highly excited while looking repeatedly along the path. Faben rushes to present his rump to Hugo, who reaches out a hand upon which Faben then bounces his scrotum. Mike displays briefly along the path while dragging the carcass behind him. There are sudden sounds higher up the path, and Charlie (adult ♂), Willy (adult ♂), Jomeo (adolescent ♂) and Sniff (juvenile ♂) appear. Raising his hair, Charlie walks rapidly toward the cluster near the carcass, then swaggers past within feet of them: Faben, Satan, and Gigi retreat slightly, apparently intimidated; Hugo stays quite still, just watching; and Mike swaggers bipedally, arms hanging downward. Charlie circles briefly, then *squeaks* as he presents his rump to Mike, who ignores the submissive gesture as he intently watches Gigi begin to tug on the carcass. Gigi jumps away with a handful of viscera. This is the first visible damage to the carcass. Mike rises and follows her, again dragging the carcass behind, and Hugo follows in turn with one hand still on the baboon. All stop after a few paces, whereupon Charlie approaches and presses his open mouth to Mike's arm. But Charlie soon turns away and charges Willy, who has been slowly approaching down the path; Willy *screams* and scrambles for a tree, and others disperse into the undergrowth. Mike appears distracted by the noise and activity: he sits near the path, motionless, hair raised, and glances about in several directions. Then, when everyone calms down again, Charlie, Hugo, and Gigi cluster around, and Mike places his free hand on Gigi's back while copulating with her. Charlie breaks away and briefly displays toward Faben, who *squeeks* and again retreats backward toward the forest.

Mike and Hugo both hold the carcass and start walking into Palm Grove at 9:03; *screaming* and *hooting,* several chimpanzees follow immediately. Faben swaggers bipedally behind the cluster that surrounds the carcass, but no one pays attention to him. Willy, Jomeo, Satan, and Sniff are last to disappear, after which there is considerable milling about in the undergrowth. Visibility is briefly impaired. Two minutes later Satan leaves the site and walks away toward the stream. Faben also departs and climbs a palm tree about 30 yards away, where he quietly begins to eat palm nuts. Showing more obvious interest in the meat, most of the others follow Mike into the trees of Palm Grove.

By 9:06 several other chimpanzees have arrived unnoticed. Hugh (adult ♂) and Pallas (adult ♀) are in the same tree as Mike, Hugo, Charlie, Willy, Sniff, and Gigi, and others are dispersed on the ground below. The carcass has apparently been partially dismembered, and portions have been obtained by several individuals. Mike retains the largest of these: the intact head, chest, and back. He and Hugo share this portion, both eating simultaneously. Charlie and Hugh possess fairly large portions, possibly arms or legs, with which they chew leaves. Jomeo varies looking up from beneath the tree with diligent searching in the grass every time fragments drop; he soon recovers a sliver of bone which he wadges[9] with leaves. Several mothers are also present: Passion (adult ♀) joins Jomeo to search for falling fragments; Mandy (adult ♀) climbs up to a male cluster, gets a piece from someone, and descends again to eat it. Willy approaches on the ground, searches, and also recovers something.

Except for occasional bursts of *pant-hooting,* sometimes in response to vocalizations up the valley, all the chimpanzees here are relatively calm and quiet. Chimpanzees are dispersed in several trees and on the ground. Perhaps because many already have pieces of meat, there are not stable clusters surrounding those in possession of large portions. Mike still has the largest portion, the forequarters, and the baboon's head remains intact. Visibility from the ground is not absolute, but no aggression has been recorded since the carcass was taken into the trees.

Two additional chimpanzees, Rix (adult ♂) and Godi (adolescent

9. A wadge is a mouthful of mixed foods which is chewed like a cud. Wadge chewing is discussed in the following chapter.

♂) arrive at 9:36.[10] Godi remains on the ground, but Rix immediately joins those in the tree, where he makes the rounds, from Hugh to Charlie to Mike, seemingly to examine the meat. Later Hugo and Gigi join Mike, forming a cluster. Hugo grasps the carcass and proceeds to rip it apart while it remains in Mike's hands. Having consumed his original portion, Charlie joins them briefly, then climbs away and is groomed by Sniff. Gigi and Hugo regularly take fragments from the carcass, and Gigi frequently *whimpers* as she begs from Mike, who does not release the forequarters but shows no possessiveness about the meat. By 10:10 much of the baboon has been consumed. In a frequently repeated sequence, Hugo steps close to Mike, puts his hand on the carcass, and pulls it gently toward himself; Gigi *grunts* softly as she watches this, then *whimpers* as she also grasps the carcass and pulls it towards herself; holding the remains by the neck, Mike ignores both. Hugo keeps pulling until he rips a piece away with his teeth or fingers; Gigi *whimpers* as Hugo steps away again, then also proceeds to bite off strips of meat or skin.

Mike begins work on the baboon's head at 10:17: gripping it with both hands, he pulls with his teeth at the skin around the eyes and mouth. Gigi and Hugo lean close and watch intently. By 10:21 the facial skin has been partly ripped away. Then Gigi bites several pieces from the neck until Mike swivels away, but she steps in front of him again to resume chewing at the neck and this time Mike ignores her.

Jomeo, Satan, and probably others on the ground seem to be losing interest in meat and are eating palm nuts and reed shoots nearby. Those in the central tree are spreading slowly onto several branches, a few still holding or chewing meat. The atmosphere is calm and relaxed until Leakey (adult ♂) arrives amidst a burst of general *pant-hooting* at 10:27. He immediately climbs into the tree and reaches a hand to Mike's lips; Mike swings away and Leakey follows. Moments later Mike, followed now by Hugo, Gigi, and Leakey, climbs down to drink from a nearby trickle of water. All move into a thicket (visibility temporarily poor) where Leakey suddenly starts *screaming,* then turns and attacks Gigi, who also *screams.* After chasing Gigi into a tree, Leakey returns to Mike; Gigi returns more slowly, sits several feet away, and watches Hugo groom Leakey, who now begs from Mike. More movement in the undergrowth follows shortly, and finally Hugo emerges with a sizable section of skin and meat. Gigi follows him and *whimpers* as she begs for some; she soon acquires

10. Presently in vicinity: MK, HG, HH, RX, FB, CH, WW, Satan, Gigi, Jomeo, Sniff, Pallas, Godi, Passion and Pom, Mandy and Midge.

a small bone and steps away to wadge it with leaves. Mike and Leakey also emerge from the thicket, cross the trickle, and sit down in a grassy clearing.

The skull is intact and still attached to folds of body skin and some vertebrae which obscure the foramen magnum. After a brief bout of unsuccessful begging from Mike, Leakey grasps the baboon's neck and *grunts* softly as he tears bits away with teeth and hands. Mike retains a firm grip on the head, briefly grooms Leakey with the other hand, and then simply watches Leakey eat. A few feet away, Gigi takes a meat fragment from Hugo's hand; then both turn to watch Mike and Leakey chew simultaneously on the carcass. When Mike finally tries to pull the head gently away, Leakey *hoo-whimpers* and holds his hand near Mike's mouth; Mike responds by grooming Leakey briefly, then continues to chew on the facial area. Leakey tears off a large fold of skin (Plate 7) which he chews a few times, pulls from his mouth for inspection, and soon replaces to chew again. Holding the head in both hands, Mike quickly scalps it by ripping back all the skin with his teeth. After licking the exposed, bloody surface he sucks out and swallows the right eye (Plate 8).

At 10:46 Mike again bites down on the skullcap with his canines, shoulders trembling with the effort. Leakey reaches for the head; Mike glances up into Leakey's face; both *hoo-whimper* briefly, then Leakey pulls away a bit of skin which he wadges with leaves. As Mike turns the skull about in his hands, a roughly circular hole of about 1.5 inches diameter is visible in the cranium. Mike pokes a thumb, then a forefinger into the aperture to extract some brain tissue which he at once wadges with fresh leaves (Plate 9).

Leakey is now watching very attentively—so much so, in fact, that he often stops chewing his section of skin. Once or twice he tries to approach the hole in the skull with an extended forefinger but each time withdraws when Mike blocks it with his own probing fingers. Nearby, Gigi has persistently and sometimes successfully begged from Hugo, who once removed a piece from his own mouth with his fingers and handed it to her. Gigi is now engrossed in wadging this with leaves, while Hugo, having finished eating, swats flies with both hands as they buzz past his face. He and Gigi are only a few paces from Mike and Leakey, yet the pairs pay little attention to one another. Others are still scattered around the area, just sitting or interacting quietly. Sniff is nearest the meat-eating cluster and constantly watches them; he keeps edging tentatively toward the cluster but hastily retreats nearly every time one of the adults looks at him.

PLATE 7

Having consumed most of the carcass, Mike (facing lens) and Leakey grasp the head and clean the skin lining by repeatedly pulling strips between their teeth.

As Mike begins work on the head, Leakey (right) continues tugging until a large fold of skin tears away.

(Top) Mike scalps the baboo skull by grasping the head both hands, one set of finge across the face, and pulling th skin over the cranium with h teeth. (Middle) After removi all skin from the head, as we as the remainder of the nec Mike licks the exposed sku and sucks out an eye. (Bottom Cupping the skull tightly both hands, Mike bears dow with his canines upon the cr nium, at the juncture of th frontal and parietal bones.

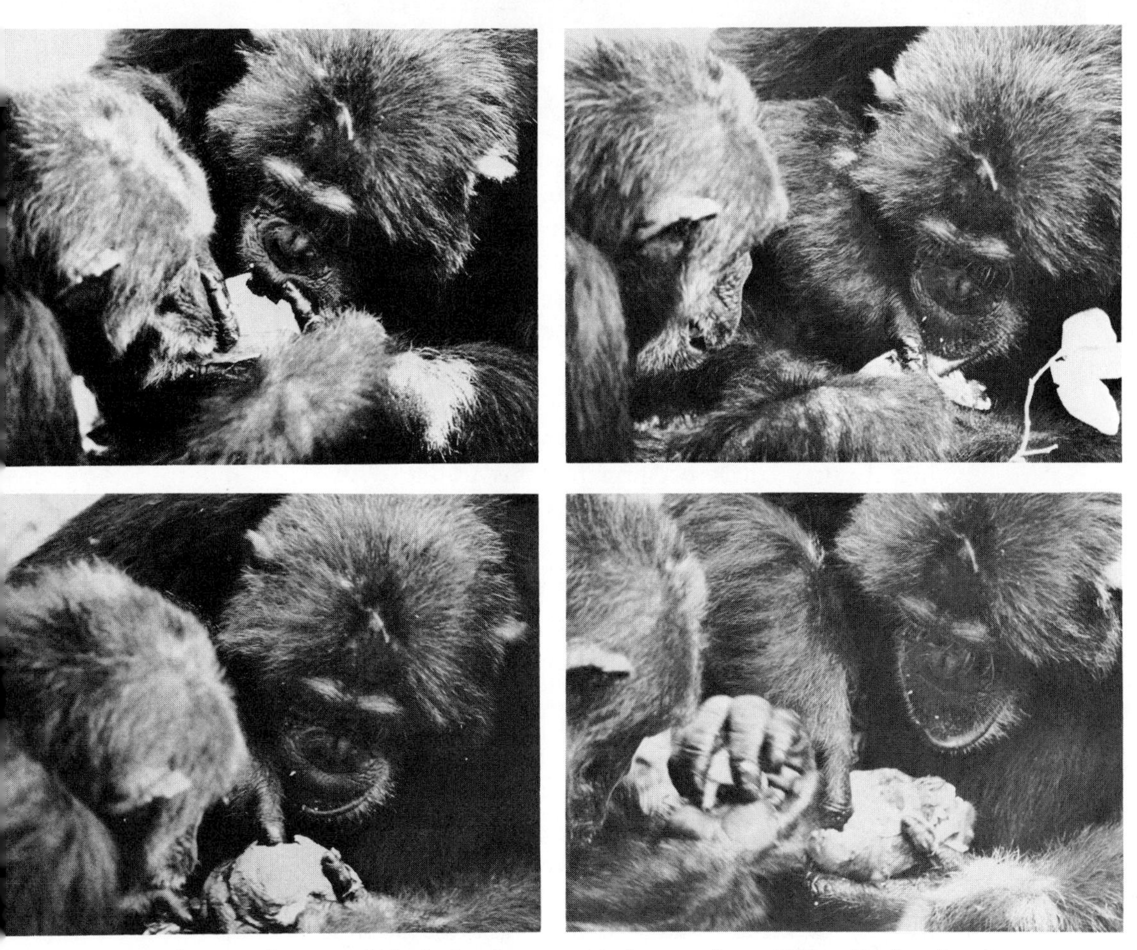

Having made a small hole in the top of the cranium, Mike (facing lens) scoops out the brain tissue by repeatedly inserting one or more fingers into the cavity; Leakey watches attentively and, still clutching a fistful of wadded skin, several times reaches tentatively toward the aperture but does not touch the skull.

Mike mixes fresh leaves with a mouthful of the brain tissue, forming a wadge; Leakey (hand visible) then begins to beg for the wadge, his fingers touching Mike's chin.

By 10:51 Mike has removed a substantial part of the brain by scraping and scooping it from the cavity with his fingers. He now removes his index finger entirely and Leakey, *panting* and shaking with excitement, finally inserts several fingers and moves them around inside. He apparently finds very little remaining, for he licks his fingers slowly and resumes chewing the fold of skin he has been holding for some time. Gigi briefly joins the males and removes something from the skull, which rests in Mike's lap while he chews a large wadge of brain and leaves. After a minute or so of steady chewing, Leakey starts begging for this wadge with one hand to Mike's lower lip and the other to his cheek (Plate 10). Mike chews a bit longer before he opens his mouth and drops out the wadge, which Leakey immediately adds to the skin already in his mouth. Inserting his forefinger into the braincase, Mike resumes scraping the interior.

Sniff, who has been watching everything from a distance of several yards, now hesitantly approaches to within a few feet of the others. Leakey and Mike suddenly turn and threaten him—Leakey arm-waving, Mike *barking* softly—but without actually leaving their places. Sniff *screams* but does not retreat. Leakey starts to lean toward him, then stops in mid-motion and turns back to resume begging from Mike. Rapidly quieting down, Sniff inches closer again and intently watches them. Gigi and Hugo sit farther away and seem to have lost interest in meat. Satan, Jomeo, Winkle (adolescent ♀), and Mandy with Midge (infant ♀) are visible from this spot, and others can still be heard in the vicinity.

By 10:58 much of the baboon's face and skull has been demolished: only part of the maxilla, the entire mandible, and the base of the braincase remain intact. Mike still bites and pulls the cranium apart, gradually enlarging the hole (Plate 11). Leakey once more arm-waves at Sniff, causing him to *whimper* and crouch away, then everyone watches Mike work. Gigi soon shifts a bit closer and retrieves a fragment of discarded bone from the ground. Hugo moves away to eat leaves nearby, and Leakey joins him shortly. As soon as Mike is alone, Gigi approaches and begins to beg, often reaching up to touch the skull as Mike takes it apart.

At 11:09 Gigi suddenly *screams* loudly as Mike stands up with hair erect, Hugo jumps up, Leakey stares, Jomeo retreats hastily, and Sniff climbs a tree. When Mike sits again, Hugo returns to groom him briefly, and Gigi returns to beg. Sniff cautiously approaches him from behind as Mike breaks sections from the skull, but someone notices, and the adults instantly turn on him: Gigi *screams,* Hugo

The brain-leaf wadge in Mike's mouth, reduced to a pulpy mass after some minutes of chewing.

Wads of skin still visible in both hands, Leakey (right) again begs for Mike's wadge by holding one hand beneath Mike's chin.

Scraping the empty cranium one final time, Mike begins to crush the skull between his jaws.

After breaking it into several large fragments, he completes the dismantling operation with both hands.

Lips pursed with concentration, estrous female Gigi (right) peers very intently as Mike finishes pulling apart the cranium.

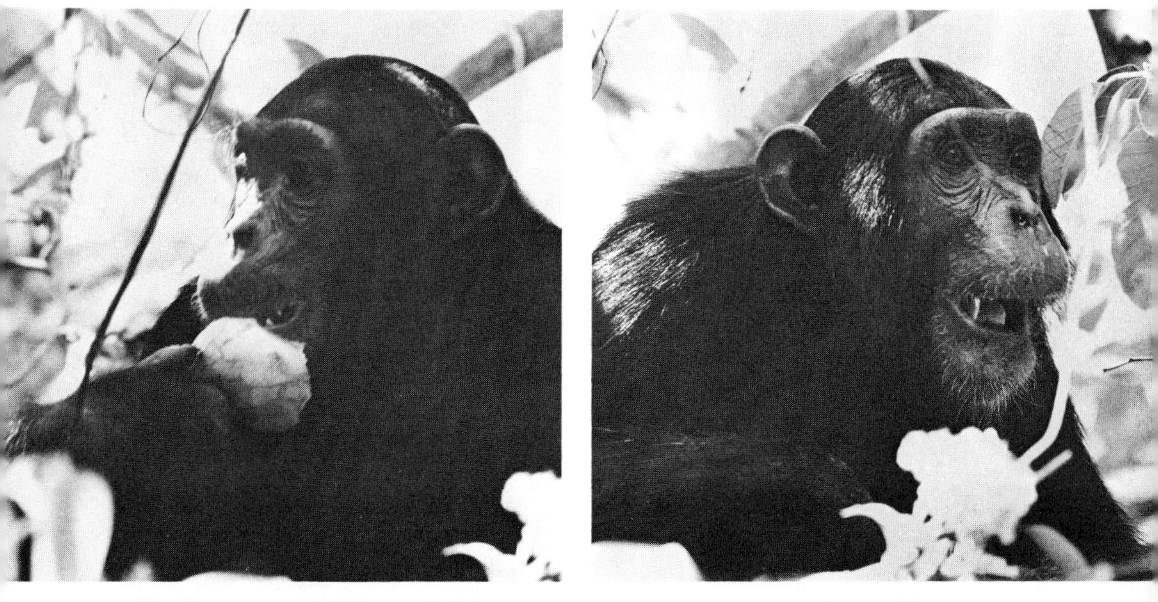

Having recovered the occipital bone from the ground and then withdrawn to a nearby tree, Sniff licks and chews it leisurely.

waas, and Leakey raises his hair and shakes a branch. Sniff *whimpers* and crouches low to the ground but again does not retreat, and a moment later he is holding a piece of skull which he takes into a nearby tree (Plate 12). The adults calm down quickly, and Hugo resumes swiping flies in the air.

Gigi is still begging persistently at 11:13 and has several times been successful. Mike now takes the mandible between his molars and tries to break it. Minutes later Hugo and Leakey move away several yards, and Mike interrupts his eating to follow a few paces. Sniff descends to the ground, approaches the vacated spot, and searches diligently until he recovers several bits of bone, skin, and meat. Gigi joins Mike and begs. When she turns away again, she is in possession of the remainder of the skull: the entire mandible and part of the maxilla. Hands now empty, Mike finishes whatever is in his mouth before lying down in the grass. Gigi bites and sucks on the mandible, then breaks off a branch and wadges leaves with it. Sitting a few feet away, Sniff chews a section of clean cranium, with other fragments of skull and a portion of an arm or leg in his groin pocket.

Gigi has all the remaining pieces, including the teeth, in her mouth by 11:33. Sniff wadges leaves nearby. Mike, Hugo, and Leakey relax some yards away, not interacting. During the ensuing half hour several chimpanzees are seen in the vicinity, some stopping with Leakey and Hugo while others simply pass by in various directions. At 12:13 P.M. Gigi chews apart the last piece of maxilla, dropping several teeth; she scratches briefly, then collects the teeth and consumes them also. Sniff, who still has a bone in his mouth and another in his groin pocket, starts playing with infant Goblin, who just arrived at the site. Several males groom nearby in a group session.

After drinking from the water trickle at 12:30, Gigi slowly wanders away and is followed by nearly all the remaining chimpanzees, including Mike, Leakey, and Hugo. Within minutes, Faben and Godi approach the spot where the brains and skull were consumed, and both search for fragments. Faben soon finds and eats a section of the skullcap. The ground is examined closely after these finally depart, and not one piece of the baboon could be found anywhere.

KILL: DECEMBER 23, 1968

This episode illustrates how several chimpanzees divide a carcass after a collective kill without overt competition. Three observers

(Staren, McGinnis, Teleki) were taking notes and photographs for 5 hours. Observing conditions were excellent, with only minor difficulties at the beginning because of rain and at the end because the chimpanzees climbed trees. This episode may be described as a seizure, but it contains some components of chasing. Refer to Diagram IX (Appendix C) for meat distribution sequence.

Feeding is completed in a relatively short time this rainy morning. Only a few full boxes remain at 9:30 A.M., an hour and a half after starting, and the last is emptied during a second brief feeding period around 10:10. Sufficient control is maintained over the distribution of bananas so that the chimpanzees stay calm, competition with baboons is minimal, and no interspecific aggression is recorded even though a sizable proportion of Camp Troop (ca. 30 baboons) is in the area most of the morning. At least half the troop has arrived at 7:08, only minutes after the first boxes fall open, but fortunately these depart within about 10 minutes; a second contingent (ca. 25) comes at 9:40 and stays until about 11:30. All the baboons eventually drift away into the valley.

After finishing the bananas, most of the chimpanzees disperse into trees peripheral to the feeding area. All of them—9 adult males,[11] 4 adult females, 4 mothers with infants, 4 adolescents, and 3 juveniles—then spend several hours in such relaxed activities as sleeping, grooming, and playing. Most of the males are in two general clusters: one consists of Hugh, Humphrey, Leakey, Worzle (all adult ♂) in trees just northwest, and another of Charlie, Mike, and Goliath (all adult ♂) in one tree on the south slope of the feeding area.

An infant or juvenile baboon starts *gecking* somewhere near the stream at 12:31 P.M. In total silence and with very deliberate movements, Charlie, then Mike and Goliath, leave their tree and walk quickly in that direction. Figan (adult ♂) puts his hair erect and hurries after them. Moments later, Sniff (juvenile ♂) runs through the feeding area. Just after these 5 disappear in the valley, Humphrey and Hugh enter the feeding area from the opposite direction and display in unison across the open north slope; they stop on the east slope and stare south intently. Seconds later chimpanzee *screams* and *waas* are heard near the stream, and Hugh, Humphrey, and Worzle hurry off toward the sounds.[12]

11. MK, GOL, LK, HH, HM, WZ, CH, FG, De.

12. Other observers later reported that even the chimpanzees northwest of the feeding area were instantly alert to the departure of the first males. How this happens remains unknown because, as in this case, there is often no noise and poor visibility.

The first observer arrives at the stream to the accompaniment of a nearby uproar at 12:34: chimpanzees *screaming* and *waaing,* and baboons *screeching* and *barking.* Baboons seem to be dispersed everywhere, running pellmell on both banks of the stream. Worzle, Humphrey, Sniff, and Figan are visible in trees, the latter *screaming* in a tantrum. Goliath displays through a tree overhanging the stream and ends by pressing his open mouth to Mike's back. Mike stares across the stream, hair slightly erect. Charlie and Hugh sit together on the ground, also looking across the stream. All the chimpanzees appear quite tense and excited, as though one attempt to kill had already been made.

Only moments later, Charlie stands bipedally and, stiffly extending both arms above his head, displays across the stream and into the vegetation near the base of Sleeping Buffalo Ridge, Hugh running close on his heels. Mike and Goliath leap to the ground and run that way also, others following. Intense noise and activity erupts among the trees ahead (just between KK1 and KK2, Map IV).

Delayed slightly by having to cross the stream, observers arrive on the scene at 12:36, about a minute after the sounds cease, and find Mike, Charlie, Goliath, and Hugh in a tight cluster at the center of a small clearing of trampled grass. They are soon joined by Worzle, who jumps excitedly about at the edge of the cluster. Several adult baboons are wandering about in the vicinity.

Glad baboon, an already dead juvenile female from Camp Troop, is being pulled in separate directions by the 5 males who, sitting hunched shoulder to shoulder, tear at her skin and appendages with hands, feet, and teeth. Many hands grip the carcass at any available point, and each chimpanzee sways and heaves with the strain of tearing part of the baboon away for himself. The cluster is so tightly packed that we rarely glimpse the carcass. Worzle suddenly pitches backward out of the ring, rapidly recovers himself, reaches one hand back into the cluster, and pulls away a fistful of viscera. Moving about 5 yards from the others, he stuffs these into his mouth with both hands.

At the same time, alternating between pulling the carcass with one or both hands, Mike puts slivers of meat into his mouth. Then, as others continue pulling appendages in many directions, Mike raises one foot to Charlie's side, the other to Hugh's shoulder and, using both legs for leverage, leans backwards as he pulls the baboon by the neck. But the carcass cannot be ripped apart, and Mike soon drops his feet to the ground again. As he leans forward to tear away more meat, Charlie briefly manages to pull the baboon close to his chest; he huddles over it, jerking his shoulders from side to side, and nearly dislodges some of the grasping hands. Mike extends both arms fully and lifts the carcass with such strength that Charlie is completely raised from the ground and Goliath nearly so. Charlie's grip slips only slightly and, when Mike relaxes again, each chimpanzee continues to pull.

Charlie and Hugh occasionally raise their heads to glance about; blood and meat flecks, mixing with rain water dripping from their hair, cover their faces and chests. Finally getting a good grip with his teeth, Charlie proceeds to jerk his head rapidly from side to side, somewhat like a terrier with a rat. Hugh also sinks his teeth into the baboon and pulls with his jaws against the opposing pressure of both hands. Goliath and Mike use mostly hands and feet, pulling with each in opposite directions in the effort to tear the skin. None of these activities have been accompanied by any direct aggression between participants.

Pulling so hard that he overturns and tumbles away from the cluster when the carcass suddenly splits, Goliath emerges in possession of the hindquarters: the rump and both legs. He recovers his balance and runs upslope into dense undergrowth. Retaining only a mouthful of meat, Hugh glances about briefly and then hurries after Goliath. Sniff, who has apparently been watching from the higher grass, *pant-bobs* submissively, and jumps away as they pass, then climbs a palm tree and watches. Charlie and Mike keep a mutual grip on the forequarters, but Mike soon manages to pull most of this section away (Plate 13). Left with only a handful of ribs and meat, Charlie *whimpers* a bit (but does not try to get the rest back) before he walks upslope after Goliath and Hugh.[13]

Leakey and Satan (adolescent ♂), who apparently has arrived during the last stages of the struggle to pull the carcass apart, remained peripheral to the others—Leakey *squeeking* with excitement and Satan wandering about quietly. But as soon as Mike is alone, at 12:40, Leakey approaches him and places a hand on the meat. Mike does not respond, even when a small piece is pulled off. Leakey then tries begging for more, one hand to Mike's chin; but Mike turns his face away and begins eating from the chest cavity. The carcass has separated at the base of the ribcage, and several ribs gleam whitely each

13. This is an excellent example of complete possession (rather than high social rank) being the decisive factor in such situations: so long as he had his hands on the carcass, Charlie had no compunctions about competing with Mike, yet Charlie immediately capitulated once Mike had sole possession.

PLATE 13

After obtaining sole possession of the forequarters, Mike briefly looks about as he holds the remains of Glad baboon, whose one eye stares vacantly from its socket.

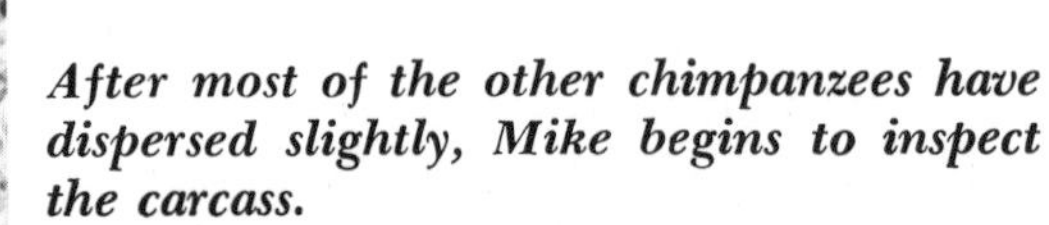

After most of the other chimpanzees have dispersed slightly, Mike begins to inspect the carcass.

Mike bites off several fingers before starting on the ribcage, some of which has been acquired by another chimpanzee.

Soon a small cluster forms: Mike wadges leaves with meat, Leakey chews and inspects the skin he obtained from Mike, and Humphrey wadges (bloody?) grass.

Pulling between teeth and hands, Mike rips away the shoulder skin to expose the underlying meat.

time Mike turns the forequarters in his hands as he selects choice morsels; the head is still undamaged, and one wide open eye stares glassily from its socket.

Then Humphrey appears at the site, *pants* and bobs in front of Mike, and joins the cluster. Neither begging nor touching the carcass, he soon turns away and collects fragments of meat or skin from the flattened grass. Frequently bending double, he also licks, rips up, and wadges grass (which may be bloody) from the spot where the baboon was recently torn apart. By 12:30 Leakey stops begging and watches Mike eat, and meanwhile repeatedly removes from his own mouth, inspects closely, and replaces the one piece he obtained earlier. After a brief interlude to lick the baboon head and bite off several fingers, Mike continues to work on the chest cavity, bit by bit breaking off ribs which he wadges with leaves (Plate 14). Humphrey leans close and removes a small piece from Leakey's mouth, again licks the moist grass for a while, and finally turns and peers closely at Mike's face but without touching either Mike or the baboon.

There is a burst of *pant-hooting* about 20 yards away in a large tree where Goliath is eating the hindquarters. Barely visible in the same tree are Charlie, Hugh, Figan, Worzle, Godi (adolescent ♂), Winkle (adolescent ♀), Fifi (adult ♀, estrous), and Gilka (juvenile ♀). As the foliage is more dense there and the observing distance much greater, the observers continue to concentrate on the cluster around Mike.

Humphrey finally reaches a hand tentatively to Mike's face, but Mike turns away at 12:56. Leakey moves closer and touches, then pulls one of the baboon's hands (Plate 15); Mike watches briefly without moving, then sneers at Leakey, who simply *hoo-whimpers* without releasing the hand. At this moment Flo (adult ♀), with Flame (infant ♀) ventral and Flint (infant ♂) behind, arrives from the valley bottom and catches the attention of the males: they look at her as she *pant-hoots* and *pant-grunts* at them but resume eating as she sits down to watch from the edge of the small clearing. When Flo hesitantly approaches the cluster moments later, *squeeking* softly, Humphrey raises his hair and stands to embrace her from behind, one arm around her waist. Mike and Leakey ignore these actions as both eat from the carcass simultaneously. Flo then sits beside Mike and watches everything intently, and Humphrey resumes licking the flattened grass nearby.

Instead of joining the cluster of adults, Flint watches briefly from a distance and then disappears upslope. He is soon spotted in the

PLATE 15

Mike (left) watches calmly as Leakey joins the meat-eating by tugging on an arm.

When Mike resumes eating from the neck, Leakey begins to chew a baboon hand.

As Mike cracks ribs between his molars, Leakey bends low and continues on the arm.

Humphrey (far right) watches intently as Mike pulls pieces from the ribcage and Leakey (facing lens) tries to bite away a hand.

Mother Flo (top) arrives, with infant Flame ventral, and joins the cluster.

Mike and Leakey ignore the others as they work intently together.

Having broken through the skin of the upper arm, Leakey tears away some of the muscles.

large tree with the second cluster of chimpanzees around Goliath, who continues to eat from the hindquarters. At least a dozen chimpanzees are dispersed in that tree, but visibility from the clearing is quite poor.

The mild rain stops at about 1:00, leaving everything dripping and much moisture in the still air. Mike, Humphrey, Leakey, and Flo sit in a tight circle around the carcass, shoulders nearly touching (Plate 16). Leakey grips the baboon while it is in Mike's hands, and both chew vigorously on the skin. Flo simply watches, neither begging nor touching anything. Humphrey tries once to beg from Mike, hand to his chin, but still seems most wary about direct contact with the carcass. He is, as a result, quite unsuccessful in obtaining anything until he places a hand beneath Leakey's chin; Leakey removes a small piece from his own mouth, and Humphrey *squeeks* excitedly as he takes it gently from Leakey's fingers (Plate 17).

Suddenly Mike rises, walks away from the cluster of chimpanzees, and climbs into a small sapling that bends nearly horizontal when Leakey and Flo follow. Mike sits with his back toward them, so neither can move close to the meat. Humphrey remains on the ground below, where he searches for and soon finds a few fragments. At 1:05 Mike turns around and, while he eats from the chest, allows Leakey to eat with him. Flo looks over Leakey's shoulder but does not try to participate. Leakey, gripping the head and shoulders of the baboon with both hands, almost pulls it all away from Mike, who retains a grip only on one of the arms. Immediately showing more interest, Flo stands up to watch Leakey bite away several ribs. Meanwhile, Mike chews slowly on a hand. Then, suddenly standing bipedally on the sapling, Mike tries to pull back the carcass from Leakey, who keeps chewing without releasing it. Soon Mike repeats the pulling—a little more strongly this time—and succeeds in partially retrieving it: the roles are reversed now, with Mike again holding the head and Leakey one arm. Flo's interest wanes again, and she descends to the ground, where she joins Humphrey in search of fragments in the grass.

Mike appears to loosen his grip slightly, and Leakey abruptly swings the carcass about so the head and shoulders are again in his possession at 1:10. Still gripping one of the arms, Mike watches calmly as Leakey strips slivers of meat from the chest cavity. Looking up from below, Humphrey and Flo start to *pant-hoot,* and Humphrey climbs the sapling to sit beside the meat-eaters, whom he watches intently. When Mike leans forward to bite vigorously into the carcass, Leakey pulls it back toward himself without hesitation. Humphrey

PLATE 17

Completely tolerant of one another, Mike (left) and Leakey even touch heads as both scrape the inner surface of the chest skin with their teeth.

(Left) Humphrey (foreground) begs briefly, but unsuccessfully, from Mike with the common hand-to-chin gesture. (Right) Humphrey (left foreground) then turns to Leakey (right foreground), who removes a fragment of meat from his own mouth; Humphrey is permitted to take this meat from Leakey's fingers.

immediately reaches for the baboon, then touches Leakey's face instead. After watching these activities, Mike grasps the baboon's head and again pulls it away from Leakey, who *whimpers* and *squeeks* as he releases it. Mike raises his hair slightly, transfers the baboon to his mouth, and leaps to the ground. Both *screaming* now, Leakey and Humphrey follow and, when Mike stops after only a few paces, Leakey embraces him from behind. Mike thereupon sits down, again in the trampled clearing, and rests the carcass on his thighs. The others immediately gather around to pick at the meat, and even Flo occasionally removes scraps.

High in the tree with the second cluster, Flint is heard *whimpering* at 1:14, perhaps because of unsuccessful begging. Flo stands excitedly at the same time, *hoo-whimpers* briefly while touching Leakey's mouth with one hand, and then resumes pulling fragments from the carcass. Humphrey leans forward suddenly, trembling with excitement, and for the first time bites into the carcass as it lies in Mike's lap. Flo releases the baboon and reaches toward Mike, who turns his face away; she moves behind him and watches over his shoulder while wadging leaves with meat. With no reaction from Mike, Humphrey appears to become increasingly bold as he chews a baboon arm. Then both chew on the ribcage simultaneously, faces only inches apart. Humphrey now and then glances up toward Leakey, who sits quietly while picking his teeth with one hand and resting the other upon the carcass. *Whimpering* softly, Flint returns to the clearing and sits at the edge; ignored even by Flo, he watches the adults intently.

Mike raises the carcass toward his mouth at 1:17, but Humphrey retains one arm and crouches forward to continue tearing at the skin. Mike briefly touches Humphrey's groin, then resumes chewing on the neck without further discouraging him. Now Leakey's attention returns to the baboon, which he begins pulling with both hands. Tugging more and more strongly, Leakey finally manages to rip away one complete arm, moves a few yards from the others to inspect it briefly, and then climbs a nearby tree where he wadges leaves with slivers of meat and bone. Flo finds a piece of bone in the grass and also moves away. Flint, who has slowly been edging toward the adults, now goes to Flo, carefully removes the clean bone from her fingers without meeting resistance, and retreats several yards into the brush to suck on it. Flo soon returns to the carcass and leans down to bite it several times in rapid succession without removing it from Mike's lap. Having remained quietly in his palm tree ever since the kill, Sniff now descends and watches the cluster from the edge of the clearing.

Much of the forequarters has by this time been consumed: the ribcage, shoulders, and one arm are gone, leaving only the intact head and part of one arm attached by a large fold of skin. Having met no resistance from Mike about chewing on the carcass, Humphrey now takes the head between his teeth and bites down several times, perhaps intending to crack the skull. Mike instantly pulls away and places the remains in his lap, covering them with one hand. Still undiscouraged, Humphrey continues reaching for the meat; but Mike shifts the carcass away several times, and finally turns bodily away. *Squeeking* loudly, Humphrey follows and again sits close to him; Mike touches Humphrey's chin, and Humphrey in turn *pants* and presses his open mouth against Mike's neck. Then both sit quietly side by side, Humphrey no longer persisting in his efforts to touch the carcass.

Flo approaches again and watches Mike resume chewing on the fingers. Humphrey resumes *squeeking* as he reaches out to touch Mike's hand, then begins to *scream* loudly while pulling at a loose fold of skin. Both tug briefly until Mike, who is starting to show signs of irritation by glaring at the others, stands bipedally and wrenches the baboon remains from Humphrey's hands. Distracted momentarily by general *pant-hoots* from the second cluster upslope, all here respond similarly. Following the vocalizations, Mike leaves the clearing and walks upslope, carrying the carcass in his mouth. Leakey climbs down from his nearby tree at 1:27 and, together with Humphrey and Flo, follows Mike toward the other cluster. As they walk through the underbrush, Mike transfers the carcass to one hand and drags it along the ground; Leakey lifts the other end to his mouth and chews as he walks (Plate 18). Sniff bobs and *pants* as the adults pass close by, then returns to the clearing alone to lick the grass and search for meat fragments.

Mike, Humphrey, and Leakey climb the tree containing Goliath and the others and instantly provoke an uproar: chimpanzees *scream,* display, and scatter in many directions through the branches. Several leap into adjacent trees and some drop into the bushes below. When Mike settles on a branch high in the tree, the others begin to calm down and regroup.[14] Goliath is observed for the first time since leaving the clearing; he appears to have consumed very little in the

14. Presently in the vicinity: MK, GOL, LK, HH, HM, WZ, CH, FG, Flo and Flame, Flint, Fifi, Godi, Gilka, Winkle, Satan, Sniff, Nova (adult ♀).

Mike holds the remains of the carcass, of which only the head and one arm are intact, while everyone responds to "pant-hoots" from another meat-eating cluster nearby.

As the cluster departs from the kill site, Leakey holds onto and eats from the carcass, even as Mike pulls it along by one hand.

past hour as he still holds two intact legs attached by strips of skin. Charlie chews a mouthful of meat which he may have obtained from Goliath.

Most members of the newly formed group have settled down to resting, grooming, and other relaxed activities by 1:40. Charlie soon finishes his mouthful and lies back along a branch. Sitting some yards from him, Hugh licks his arms thoroughly (collecting moisture, either rain or blood) before climbing over to join Mike, Flo, and Fifi, who have formed a small cluster in the meantime. Humphrey, who has climbed straight to Goliath after entering the tree, places one hand on the hindquarters but does not attempt to remove anything. These constitute the second, even smaller cluster of meat-eaters. Figan, Worzle and others repeatedly interrupt their various activities to look down at the cluster surrounding Mike, but none moves closer. Sniff stops his search of the clearing, approaches the large group, and climbs a nearby tree with a small piece of clean bone protruding from his lips.

Occasional baboon vocalizations continue lower in the valley near the base of Sleeping Buffalo Ridge, so at least some of the troop are still in the area. None has, however, been seen since shortly after the kill. There have also been sporadic bursts of *pant-hooting* from other chimpanzees higher on the ridge, but those at the meat-eating site ignore most of the calls.

Just after Fifi and Hugh climb away from Mike, neither having tried to beg or take meat, Leakey and Charlie join Mike and Flo. Having finished the arm he obtained earlier from Mike, Leakey touches the baboon head several times but takes nothing. Flint approaches the cluster and stands just behind Flo, chewing his bone fragment while watching Leakey. Quite inactive for the moment, Flo does not even watch consistently.

While Mike's cluster is thus occupied, Goliath leaps to an adjacent tree, and Humphrey follows alone but still makes no attempt to share the meat. Worzle joins them at 1:54 and, when he turns around only moments later, holds a piece of meat which he must have begged or taken from Goliath. Humphrey calmly watches Worzle consume this.

With Flo watching more closely again, Mike finally starts biting into the baboon skull at 2:00 in his first attempt to break it open for the brain.[15] Leakey quickly touches the head but withdraws his

15. Although chimpanzees have been observed to extract brain tissue through the foramen magnum, no effort to do so was made here although this opening was unobstructed. It is thus possible that some preference exists for breaking through the cranium, even in cases where another route may be easier.

hand as Mike turns his back toward everyone. Charlie soon stops watching, self-grooms briefly, and departs to lie down on another branch. Mike shortly climbs higher in the tree, finds a new seat, and resumes working diligently on the head. Only Flo follows and watches intently. The view is temporarily obstructed by foliage and no details of the entry into the skull are seen. At 2:15 Hugh and Leakey climb up to sit near Mike, who now wadges leaves—presumably with brain tissue—while holding the skull with his feet. An opening of between 1 and 2 inches is visible at the top of the cranium, which has not been scalped entirely.

In the meantime Humphrey has finally started to beg from Goliath, but without success. Now Nova (adult ♀) joins them, presents her side to Humphrey, and is briefly groomed. Goliath turns and looks at them, then continues chewing leisurely. When the grooming stops both Nova and Humphrey turn to look at Goliath; soon Nova climbs away, and Humphrey begins to self-groom sporadically. These activities continue for several minutes, but at 2:10 Humphrey suddenly displays for some unknown reason at Nova, who *screams* and scrambles away. Several chimpanzees turn to watch as both drop from the tree onto the steep hillside, where Humphrey chases her upslope while slapping her back, drags her downhill by the neck, slaps her some more, and finally displays away into the undergrowth amidst the sounds of tearing shrubbery and cracking bushes. Nova eventually stops *screaming* at 2:16, and Humphrey returns to Goliath, *pants,* and sits beside him. Soon Nova climbs the tree also but sits some yards higher with Worzle.

Minutes after Humphrey calms down, Fifi approaches him, presents, and they copulate. Afterward she leaps away to another branch. Humphrey again seats himself at Goliath's side yet makes no attempt to obtain meat from him. Mike is still obscured by foliage, so the state of the skull cannot be determined.

Following a quiet period during which several chimpanzees relax or self-groom, the entire tree suddenly erupts with *screams, waas,* and frantic movement at 2:38. Perhaps taking advantage of the general disturbance, Figan swings and then runs toward Goliath across several branches and moments later reappears with a handful of meat. Apparently Figan took him by surprise and snatched a piece while moving at considerable speed. Goliath instantly stands bipedally and *screams* toward Figan's receding back but does not pursue. Instead Humphrey launches himself into an elaborate display through the upper portion of the tree, then drops to the ground and disappears from view as he slaps and stamps downhill through the vegetation. Also *screaming,* Leakey reaches toward Hugh, who pats the upturned palm repeatedly. Having leapt into another tree at the outset, Mike now displays back through the other chimpanzees, causing additional activity and *screaming.* Other displays follow, but details cannot be seen from below.

Everyone has again settled down by 2:40, but only briefly. Sitting alone in a low fork some yards from the nearest chimpanzee, Figan pauses and rapidly finishes eating his portion of meat. He then runs along a branch toward Godi, who *screams* and drops into the bushes below. Figan stops, but Charlie raises his hair, leaps after Godi, and chases him briefly along the ground. Mike again displays through the tree, skull in mouth, and causes several others to disperse rapidly. By the time everyone stops climbing at about 2:44, Goliath and Mike are in opposite sections of the tree, both still in possession of their meat. Goliath and Leakey sit together near Hugh, who ignores them both; Leakey merely watches Goliath eat without actively participating. Both legs of the baboon seem essentially intact, so the nature and source of Figan's piece remain unclear. Mike settles down with only Flo nearby; although he seems to have consumed the brain, he continues to lick the interior of the cranium.

The ensuing half hour passes slowly; none of the chimpanzees moves about, not even those having meat show much interest in eating, and only infrequent *pant-hoots* higher on Sleeping Buffalo Ridge break the silence of the valley. For some time Mike continues working leisurely on the skull, sucking and chewing the bones, but he is too high for detailed observations. Only Flo stays constantly with him, watching and sometimes begging. Leakey joins the cluster at 3:50, displaces Flo from her position next to Mike, and starts to beg with a hand near Mike's face.

Meanwhile Goliath has climbed to a branch just above Mike, where Charlie and Worzle now join him in a second cluster. Figan sits a few yards from them yet rarely glances toward the three. No other chimpanzees are showing interest in meat. Goliath does not eat the meat he holds: he chews what is already in his mouth very slowly and sometimes picks at the rest of the hindquarters. After grooming Worzle several minutes, Goliath shifts to another branch. Worzle follows, *hoo-whimpering,* then reaches one hand to Goliath's lips and places the other on the baboon legs. He seems to have developed a sudden interest in the meat (which may be related to Goliath's disinterest), as he repeats these actions 6 times—unsuccessfully—before Goliath finally turns and climbs away. Worzle sits alone a few seconds, then follows and resumes begging with even more persistence. Goliath at first ignores these advances, but he eventually threatens Worzle with a soft *bark* and an arm-wave at 3:29. Worzle rapidly retreats a few paces and is immediately replaced by Humphrey, who begins to self-groom while sitting beside Goliath.

A single female baboon—probably Glad's mother—descends from a nearby palm tree at 3:44 and disappears in the direction taken earlier by Camp Troop. The chimpanzees pay no attention to her.

The group of chimpanzees still under observation at 4:00 P.M. are in 2 general clusters around those holding meat: one cluster centers around Goliath, with Humphrey closest, then Figan and Nova; another focuses on Mike, with Leakey sitting beside him and Flo and Fifi a bit farther. Flint, Charlie, and Gilka are dispersed in the tree, all resting. One or another occasionally climbs to a new perch, and some groom sporadically. Only Leakey, Flo, and Fifi maintain partial interest in the remaining meat. At the moment all 3 beg persistently from Mike, though without success. The other chimpanzees are slowly leaving in many directions. At 4:16 chimpanzees *hoot* and drum on trees high up in the nearby ravine (KK 1), Hugh among them.

Fifi's persistent begging is finally rewarded at 4:21 when Mike permits her to take from his hand the last small bit of skull. Climbing immediately to another branch, she sucks and chews on this diligently. Leakey begs a bit longer for whatever remains in Mike's mouth, somehow obtains a fragment, and also climbs away to eat.

Flo has meanwhile joined Goliath and Humphrey. Her persistent but unproductive begging is interrupted at 4:27 because Humphrey raises his hair, leaps to her, and violently shakes the branch while pressing her against it with his chest. He then steps back to Goliath, who suddenly displays away across several branches with Humphrey following suit. Several chimpanzees *scream* and scramble out of the way. A few minutes after calm settles again, Flo slowly climbs back toward Goliath; Humphrey stands and raises his hair as he glares at her; Flo backs away slightly and sits down; Humphrey resumes his position beside Goliath, who ignores these activities and continues to chew slowly.

Only Mike, Fifi, Humphrey, Goliath, Nova, and Flo, with Flame ventral, remain in the tree by 5:00; Flint is some distance from the adults. Goliath still holds one intact baboon leg. After 3 hours of being with Goliath without attempting to get meat, Humphrey takes a small piece from Goliath's mouth at 5:12. Having by this time consumed the skull, Fifi approaches them again and lies down nearby.

Flo soon joins the cluster, this time without producing a reaction from Humphrey, and watches the males chew. Slowly shifting the last bit of meat about in his mouth, Mike stretches out along a horizontal branch. Nova begs persistently but unsuccessfully for this piece with one hand to his lips, then moves away when Mike swallows for the last time.

The last observer leaves at 5:35. Only Goliath has meat: the lower portion of one leg. Humphrey, Flo with Flame, Fifi, and Flint stay nearby, but none has attempted to obtain meat for some time. Alone in another part of the tree, Mike reclines on his back. On the way back to the feeding area a thorough search of the kill site yields nothing, not even small fragments of bone or skin.

At 7:00 in the evening, with the light fading rapidly, Goliath, Leakey, and Godi come to the feeding area. As they pass through on their way to nesting sites on Peak Ridge, Goliath still has a bit of meat in his mouth but none in his hands. Six and a half hours have passed while the chimpanzees consumed one small juvenile baboon.

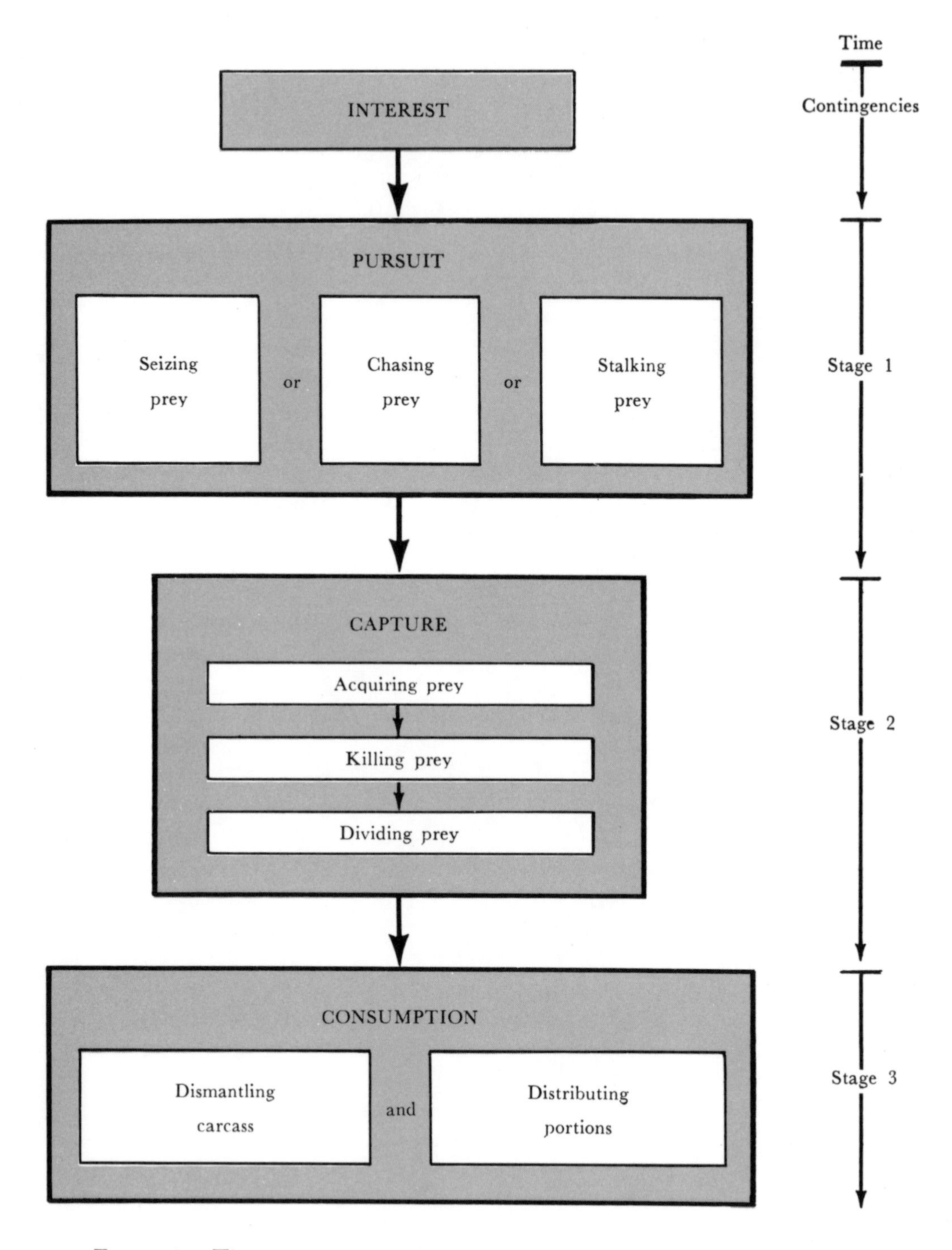

Figure 2. The major stages and components of predatory behavior that occur in complete episodes.

3

THE STAGES OF PREDATION

Chimpanzee predatory behavior consists of consecutive activity stages: pursuit, capture, and consumption. Although there is variation in episodes, all of these stages are always present in successful predation. Each stage contains discrete behavioral elements and patterns that are absent in other stages (fig. 2).

The first stage, *pursuit,* is separable into different modes: (1) seizing, (2) chasing, (3) stalking prey. Mode–1 may broadly be characterized as "opportunistic," because chance events and circumstances—e.g. unexpected discovery of prey—are of prime importance. Seizures tend to be explosive acts lacking preliminaries, and some seizures are probably analogous to the reports on prey seizure by savanna baboons (DeVore and Washburn, 1963; Altmann and Altmann, 1970). Mode–2 and mode–3 include the search and selection of prey, which are the critical elements of pursuit. Distinctions between these modes tend, however, to be more a matter of degree than kind. As true omnivores, the Gombe chimpanzees can practice predation optionally and can apparently choose from several styles of pursuit. Subsequently, they are confined neither to the irregularity of the "opportunistic" meat-eater nor to the regularity of the specialized carnivore.

The second stage of predation, *capture,* contains sequential substages: (1) acquisition, (2) killing, (3) division. The first refers to the manner and circumstances in which prey are taken or grabbed, the second to the techniques used in dispatching captives, and the third to the manner in which the carcass is initially dismembered among the captors.

The final stage, *consumption,* includes primarily the sequence in which carcasses are commonly eaten and the procedures whereby meat is distributed within the attending group.

Predatory episodes are sometimes preceded or introduced by special factors and conditions that may stimulate pursuit. These contingencies are at times clearly recognizable, but more often they are elusive even for experienced observers of chimpanzee behavior. Description and analysis of the diverse contingencies to predation and the activity stages comprising predatory episodes form the basis of this chapter. Information from all 30 episodes recorded in the 1968–69 period, as well as pertinent material from earlier and later episodes, is used to sketch the currently familiar but still incomplete picture of this special behavior in the Gombe chimpanzees. Numerical analysis is limited to the study period between March 1968 and March 1969. As with all composite sketches of primate behavior, there arises the problem of creating an overly generalized abstraction of reality, and the chapter should be read with this in mind.

FACTORS AFFECTING PREDATION

Systematic and reliable assessment of the many factors which may elicit and affect predatory behavior has many hazards. On the one hand, episodes where the preceding situation was documented in detail are rare, so the available episodes appear to contain many unrelated variables. On the other hand, very little is known about the local numbers and habits of most potential prey species. Nevertheless, some consideration must be given to the broad categories of factors upon which predation may be contingent.

Because there is such a large amount of data on predation by chimpanzees from the Gombe Research Centre in comparison to other field projects on nonhuman primates, possible relationships between banana feeding and predatory behavior must be examined. More specifically, several points are of interest and concern: To what extent might the decade of predation observed at Gombe have been the result, directly or indirectly, of the procedure of providing bananas to chimpanzees? And how might regular banana feeding (up to June 1968) or the conditions arising therefrom have affected predation between March 1968 and March 1969? If there are grounds to suspect that predation was unusually frequent during the 1968–69 period, then it would also be interesting to learn whether certain prey were more available at that time, whether chimpanzees learned the behavior or its rewards more rapidly, or whether observers became more efficient and thorough in gathering this kind of data.

The reply to the first question would, on numerous grounds, appear to be negative. Predatory behavior was documented in the very early stages of the Gombe research program, prior to the introduction of regular feeding (Goodall, 1963a). Unfamiliar chimpanzees not exposed to bananas have several times been observed eating meat outside the core study area. Chimpanzees of the study community did kill prey on days when bananas were not being distributed, and in places far from the feeding area in Kakombe Valley. Moreover, predatory behavior has been observed in 2 populations of chimpanzees that were not being provided with food—one in Kasakati Basin (Kawabe, 1966) and another in Budongo Forest (Suzuki, 1970). It is unlikely, therefore, that human intervention has resulted in the emergence of predatory behavior among the Gombe chimpanzees.

The second question is more difficult to assess properly because the many developments in research procedures over the past decade at Gombe cannot be reversed for reexamination. However, a tentative answer is again negative, though not an unqualified negative. The high predation rate observed between mid-March and mid-June of 1968, 14 cases in 3 months, seemed at the time to correlate with the large number of chimpanzees and baboons massed at the feeding area. The reorganization of research procedures in June 1968 was an attempt to reestablish more natural behavior patterns, and the severe reduction in provisioning from 7 to 1 or 2 banana days every 2 weeks was, among other changes, expected to directly influence baboon attendance and, thereby, to affect predation upon them. These expectations indeed seemed validated in subsequent months, for the rate of baboon attendance dropped sharply and the rate of predation fell correspondingly between mid-June and mid-November to only 5 cases in 5 months. But this period of reduced activity was then followed, between mid-November and mid-March (1969), by a new increase of 11 cases in 4 months but without a similar increase in the feeding rate (fig. 3). Perhaps even more importantly, whereas most predatory episodes in the first 3-month period occurred on feeding days, most of those in the final 4 months were not on feeding days. The temporary simultaneous drop of feeding and predation rates in June may have been quite coincidental, or perhaps part of a cyclic variation (e.g. seasonal prey availability). Several unrelated facts further support a largely negative answer to the second question. The 12 kills recorded in 1968–69 are in reasonable accord with the general average of 10 kills per year for the decade. And finally, there are 5 known prey species other than baboons, and predation on none of these was likely to be affected by banana feeding since they did not come to the feeding area.

There is, on the other hand, some evidence indicating that banana provisioning may have caused not so much a change in the frequency of predation as a marked shift in emphasis to baboon prey. Certainly baboon and chimpanzee association was steadily increased—until June 1968—by the regular availability of large quantities of bananas, and this presumably increased opportunities for predation correspondingly. Data collected after the changes in feeding procedure were completed strongly support this premise: baboons constituted 83% of the prey captured in the 1968–69 period but only 22% of those taken in the following 18 months.

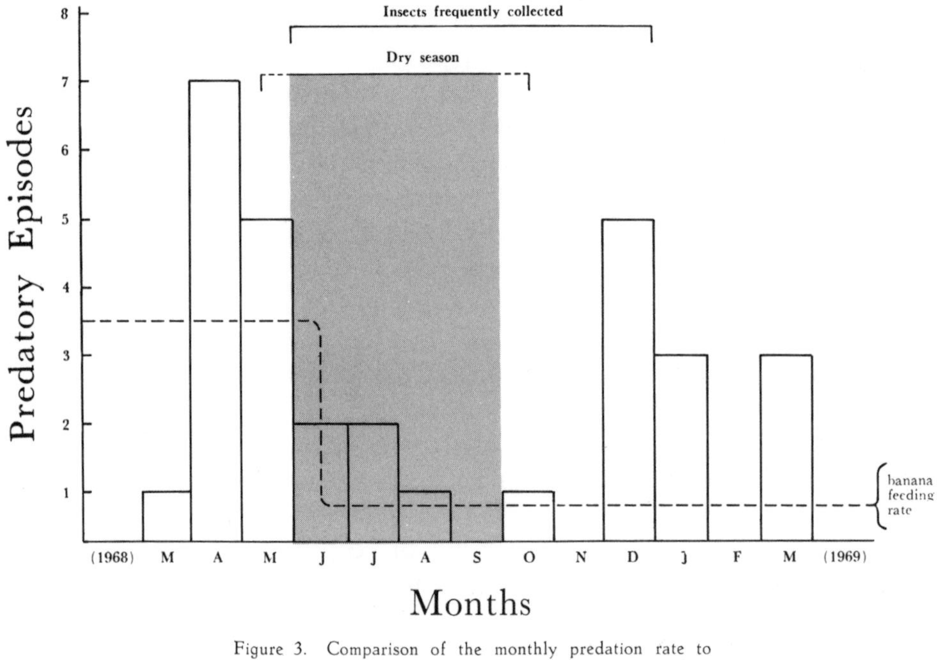

Figure 3. Comparison of the monthly predation rate to the relative drop in banana feeding (from 3.5 to < 1 times per week) between March 1968 and March 1969.

Without special study of variables that include the observers themselves, no amount of speculation can solve the dilemma of a *true* increase in predation resulting from chimpanzees learning a behavior pattern or becoming habituated to meat, and an *apparent* increase in predation resulting from more consistent, more efficient observation by larger numbers of human beings. The simple fact that more episodes of unsuccessful predation were observed in the 1968–69 period than at any other time during the decade probably indicates that observer experience and efficiency does play a significant role in the general picture.

Turning to a somewhat tangential aspect of banana feeding, available data suggest that predation does not correlate with degrees of hunger. During 1968–69 prey were pursued on nonfeeding as well as feeding days, and episodes occurred before, during, and after bananas were eaten. In the 19 cases which occurred after feeding, the average time lapse between banana-eating and initiation of predation was less than an hour. In several cases those who initiated predation had consumed more than 20 bananas. Since eating bananas is not known to stimulate an urge for meat, it would seem reasonable to assume that predatory behavior has more functions than those of nutrition.

Other factors which presumably affect predation are the availability, the size, and the behavior of various prey. First consideration might be given to the distribution and location of potential prey species. A general comparison of the grouping tendencies of 6 species shows that the colobus monkeys and baboons are the 2 mainly communal prey species (Table 4). These species comprise 65% of the 56 prey identified in 10 years. This suggests that large groups of prey are selected by chimpanzee predators, perhaps because large concentrations of prey are easier to locate and provide better opportunities for captures. Prey species that tend to be solitary or in small groups, such as blue and redtail monkeys and bushbucks, appear to be less common in the park and are more likely encountered by chance.

TABLE 4

Relative Estimate of the Grouping Habits of Potential Prey Species in Gombe National Park

Prey Species	Primarily Solitary	Small Groups	Primarily Communal	Interspecific Association*
Olive baboons		X	X	
Colobus monkey			X	
Redtail monkey*	X	X		X
Blue monkey*	X	X		X
Bush pig		X	X?	
Bushbuck	X	X		

* Redtails and blues are regularly seen in association with colobus troops, sometimes for days and perhaps weeks at a time.

Since colobus monkeys and baboons have in common fairly cohesive communities which usually range as units, one might expect these species to be preyed upon in similar fashion by chimpanzees. But colobus monkeys, like redtails, are much more arboreal than baboons. The added factor that baboons and chimpanzees share terrestrial habits[1] and omnivorous diets must also affect predation, if only because these species would tend to share locations more regularly. In contrast to predation upon baboons, which occurs mostly on the ground, predation upon the smaller monkeys—colobus, redtail, and blue—usually occurs above ground, where these prey are probably more elusive in the multidimensional forest canopy, and especially at the edges of the zone where branches are both thin and pliant. Even the collective pursuit practiced by clusters of chimpanzees is likely to provide only partial compensation for the disadvantages of greater weight and reduced mobility in the treetops.

Although all the known prey species have often been seen within the core study area of three valleys, only baboons and bush pigs are commonly encountered in the lower portion of Kakombe Valley where most episodes have been observed. However, both bush pigs and bushbucks are probably less available as prey than the various primates because they are partly nocturnal (Leakey, 1969). Combination of these factors again leads to an emphasis on baboon prey.

Although the list of 56 captured prey shows that baboons are preyed upon most frequently; colobus monkeys, bush pigs, and bushbucks less so; and blue and redtail monkeys least; there are insufficient data to firmly establish causes. Differences in predation rates upon various species are probably susceptible to a large number of unknown variables. Only among chimpanzees and baboons has the predator-prey relationship been consistently observed to date, so discussion of specific factors must at present be limited largely to this one combination of species.

One of the important factors which appears to strongly influence predation is the size of potential prey individuals. Chimpanzees have not been observed to capture individuals of comparable or larger size than themselves.[2] In fact, the size difference between predator and

---

1. Terrestrial locomotion is much preferred by the Gombe chimpanzees, even in the most rugged terrain and very dense vegetation. Arboreal activities such as resting, grooming, and feeding are essentially stationary, and chimpanzees are clearly terrestrial if motion, not relative time, is accepted as the important criterion.

2. On September 15, 1971, an infant chimpanzee, aged 1.5 years, was killed and partly eaten by adult male chimpanzees in Kakombe Valley. Neither the infant nor its mother were members of the study community. Parts of the legs were consumed, but the body was abandoned. This unique episode will be the subject of a special detailed report by the observers.

prey is always considerable: although accurate measurements are lacking, prey are usually within a 20-pound maximum. Thus, prey individuals at the higher extreme are no larger than adult or nearly adult colobus, redtail, and blue monkeys, and infant or newborn bush pigs and bushbucks. Baboons are also selected within this size range: individuals pursued during 1968–69, in the period when an accurate baboon census was kept by the Ransoms, were between the ages of 1–75 weeks, with an average close to 20 weeks (Plate 19; also see Appendix B, Table III). Only in the December 3 episode was an adult baboon held briefly during predation, and the intentions of the chimpanzees were doubtful.[3]

Another contributing factor to be examined is the awareness of the predator for the prey. Limits to distance are probably highly variable, being determined by vegetation and topographic conditions. Pursuit is generally limited to a small area, but chimpanzees have been observed to travel as much as 300 to 400 yards from the moment of showing predatory interest. Some episodes did appear to be spontaneously initiated but human limitation in signal reception was a problem. The importance of visual and/or vocal stimuli from potential prey can be illustrated by the following figures compiled from 28 events in 1968–69: (a) in 10 cases, or 35%, predation followed visual and vocal contact between predator and prey; (b) in 7 cases, or 25%, the prey was visible but not audible; (c) in another 7 cases the prey was audible but not visible; and (d) in only 4 cases, or 15%, were both stimuli apparently lacking. Fully 85% of the 28 episodes involved visual and/or vocal contact at the time of initiation, and 60% of the episodes seemed to be stimulated by prey vocalization alone. As the few episodes where all stimuli seemed absent are perhaps accounted for by better visibility from trees or greater acuity in chimpanzee hearing, predation upon baboons may correlate highly with different kinds of prey activity. The stimuli involved in predation upon smaller monkeys, pigs, and antelopes remain uncertain but are probably mainly visual.

In the 17 cases where baboon prey vocalized during 1968–69, the sounds were of 2 types—one distressive, the other aggressive—made by infants or juveniles. The most common stimulant vocalization was the *screech* (*yak* or *chirp–click* according to DeVore and Hall) of black or transitional infants in distress. Adult male baboons in both troops at Gombe often took small infants from their mothers, then

3. On this occasion Mike was seen briefly holding an adult female baboon (possibly the mother of a juvenile just captured by Hugo) to the ground with both hands. She soon pulled free and scrambled away unhurt.

Mother baboon Arwen, with her black infant of four weeks—common prey for chimpanzees at this age.

A small mixed cluster of baboons—an adult male, adult female, and black infant—typical of those preyed upon by chimpanzees.

carried, handled, or groomed them for long periods of time. Since this handling was sometimes rough as well as prolonged, distress sounds often accompanied the struggles of such infants to return to their mothers, who usually stayed nearby without offering assistance. The second type was the more aggressive *geck* (*shrill bark*) of juveniles. These aggressive calls occurred in a variety of circumstances, such as play between juveniles.

Interestingly, in the 28 cases of predation upon young baboons during the year, prey individuals were 7 times with their mothers, 5 times in possession of an adult male, and 4 times alone, with 12 cases remaining undetermined. These figures indicate that the position of a young baboon relative to adults of the troop is probably not a significant factor in prey selection, and there is also some possibility that the special bonds reported between adult male and immature baboons (Ransom and Ransom, in press) do not ensure greater safety for the young.

Disposition of the predator (s) is also likely to affect predation, although the mechanisms operating here remain very unclear. In the 24 cases in 1968–69 where chimpanzees were under observation prior to an episode, many were started when the "predators" had been relaxing and were seemingly disinterested in obtaining meat.[4] For example, feeding periods—and especially those where each chimpanzee consumed as many bananas as desired—were usually followed by relaxation periods which could last several hours (Plate 20). These were often the very times when predation was initiated. Conversely, periods of intense activity, such as the excitement precipitated by feeding, were rarely interrupted or accompanied by predation even when many chimpanzees and baboons were together at the feeding area. Moreover, the general concept that develops while observing chimpanzees and baboons in association day after day is that predation occurs rarely in comparison to the number of times prey are available. Thus, if predatory behavior is prompted neither by convenience nor by hunger, then the disposition of certain chimpanzees may be an important factor.

Daily weather does not seem related to predatory behavior, for episodes have been recorded on both clear and rainy days with no indication of pattern. However, no heavy thunderstorm has to date been accompanied by predation. The predation data collected in the final 30 months of the decade may, on the other hand, show some

---

4. Relaxation here refers to low intensity social or autonomic activity, such as play or grooming, and to inactivity, such as dozing or sleeping.

Leakey relaxes after consuming about 25 bananas—a common pastime following feeding, but also the time when predation often occurs.

Cluster grooming—here Rix on Mike on Goliath on Leakey on Pepe—is another favored activity after bananas have been consumed.

correlation to climatic variation (see fig. 1). Additional years of detailed information may yield specific correlations with other seasonal factors, such as yearly fluctuations in the availability of other protein foods. Regular consumption of galls, ants, and termites occurs in general between the months of June and January (Lawick-Goodall, 1968b), a period which roughly overlaps the months when predation occurs least (see fig. 3). Other foods consumed by chimpanzees in Gombe National Park, including many kinds of fruits and leaves, are available in seasonal cycles, and the concentration of predatory episodes in particular months may simply conform to this shifting dietary pattern.

The daily activity cycle of Gombe chimpanzees follows a general pattern perhaps best reflected in the time spent in food-getting activities. Chimpanzees leave their nests at about 6:00 A.M., exchanging vocalizations frequently as they forage or travel to feeding sites; they then feed heavily between about 7:00 and 10:00, and usually rest or participate in relaxed social activities during the midday heat between 11:00 and 2:00 P.M.; they again feed extensively between about 4:00 and 6:00, with considerable traveling and vocalizing, and settle down in nests any time between 6:00 and 7:30 P.M. During the 1968–69 period, the initiation of predation tended to coincide with the early

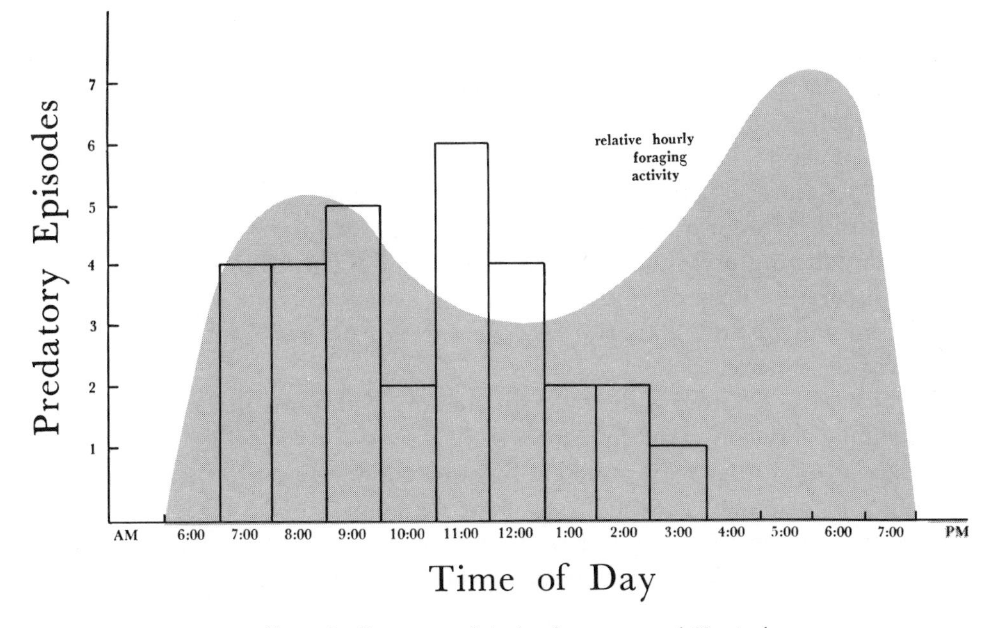

Figure 4. Comparison of the hourly occurrence of 30 episodes observed in 1968-69 and the hourly rate of general foraging activity (the latter adapted from Lawick-Goodall, 1968b:183).

portion of this daily activity cycle. Most instances of predation occurred in the morning hours of greatest general activity and tapered off sharply during midday and early afternoon when temperatures were highest (fig. 4). Very few episodes have, in fact, been observed during the second intensive feeding period throughout the 10 research years. Even those chimpanzees who ate large portions of meat obtained late in the day would sometimes forage for vegetable or fruit foods before nesting. In 3 cases observed in 1968–69, chimpanzees who still had carcass fragments in the evening carried these as they foraged, then took them into their nests. On several other occasions individuals were seen with meat that had been kept in the nest and partly eaten overnight. Night feeding on other items has also been observed.

UNSUCCESSFUL PREDATION

Only 37 unsuccessful episodes have been observed at Gombe in 10 years. Of the 18 attempts upon baboons in 1968–69 (64% rate of failure), only 13 were described in detail because 2 were recorded only on the bases of general activity and vocalizations and 3 were impossible to see clearly on account of thick vegetation and rapid movement.[5] If attempts are to be defined as the unsuccessfully terminated pursuit of prey, then lack of success may be associated with several categories of variables.

One factor that may terminate pursuit is the sudden, often unaccountable loss or change of motivation among participant chimpanzees. In spite of definite predatory intent, chimpanzees have been observed several times to stop pursuit suddenly and resume former activities. Moreover, periods of interest and disinterest sometimes alternate during prolonged stalking to such an extent that pursuit slowly disintegrates. Loss of interest does not always correlate with visible circumstances and may at times derive simply from anticipation of eventual failure.

Other, more observable determinants of failure are adverse environmental conditions that interfere with activity or visibility. The difficulty of pursuing small monkeys through the forest canopy has already been mentioned. In the cases with baboons, thick vegetation apparently aided the escape of prey on 9 occasions (70%). Baboons did not always flee from chimpanzees who were intent upon predation;

5. The 2 episodes: December 29, 1968 (the second of the day), and March 8, 1969. The 3 episodes: May 6 and 8, 1968, and March 6, 1969.

but, once intimidated into flight, the direction taken often led to higher grass or thicker brush where the pursuers appeared to have difficulty locating the prey.[6] Even with collective pursuit well underway, chimpanzees operating at ground level where the vegetation was higher than their line of vision appeared to be handicapped in locating the prey individual, particularly with the added confusion caused by many baboons milling about in the vicinity. Although failures were more common in dense vegetation than in clearings or beneath gallery forest, it is uncertain whether baboons actually used such conditions with intent. Baboons at Gombe rarely used trees as routes of escape from pursuing chimpanzees; in fact, treed baboons usually tried to reach the ground immediately even when this meant a long and dangerous leap. As savanna baboons in other regions of Africa tend to seek arboreal sanctuary from many predators (DeVore and Washburn, 1963; Altmann and Altmann, 1970), the preference for terrestrial evasion at Gombe may be related to the arboreal agility of predator chimpanzees. In Kenya, similar flight to and on the ground was observed by the Altmanns only when Masai tribesmen approached the troop. Also of interest is the fact that chimpanzees who lose sight of prey apparently prefer searching through the undergrowth to climbing trees for spotting purposes. Cooperation probably does not extend to the point of providing aid (signaling) in relocating prey. Nonetheless, tree locations do at times have an important function in the initial sighting of prey, particularly of bush pigs and bushbucks. Bipedalism is also rarely adopted as a means of seeing better during predation, yet this stance and mode of locomotion occur often in other contexts: when searching for a companion or carrying food, and in many social interactions (Lawick-Goodall, 1968b).

A third cause of failure can be evasion by the prey itself. This tactic is not always possible because of the distribution of chimpanzees during collective predation and because chimpanzees are adapted to all dimensions of the habitat. Moreover, baboons at Gombe tend not to flee immediately unless the prey individual, or the adult carrying the endangered infant, is effectively separated from the other members of the troop. In only 3 attempts (23%) during 1968–69 was flight the initial response, and the adult in each of these cases was a female more than a dozen yards away from other baboons.[7] In the remaining

6. In most open areas, grass is much higher (up to 14 feet) than a standing chimpanzee, except for some weeks late in each dry season. The feeding area was trimmed regularly to facilitate observation, but the normally tall grass and brush beyond the perimeter interfered with predation on several occasions.

7. April 27 and July 14, 1968, and January 25, 1969.

10 attempts (77%) protection was more directly interactive with the chimpanzees. When in the company of adult male baboons, females (usually the mothers) appear more inclined to stand their ground and threaten the approaching chimpanzees. However, if the distance between predator and prey is reduced below a certain limit, the female may break away and the male (s) may then attempt to delay the pursuit. But a single male baboon is not always capable of intimidating or diverting more than 1 or 2 chimpanzees.[8]

A fourth determinant of failure can be protective retaliation by the prey or, in the case of baboons and colobus monkeys, by other members of the community. Little is known about the retaliatory behavior of bushbucks and bush pigs because the pursuit stage of predation has not been observed in detail. Adults of both species have been seen to stay near the base of a tree in which their young were being consumed and to repeatedly charge subadult chimpanzees who tried to descend for fallen scraps. Among the smaller monkeys one common response is collective harassment; in nonpredatory situations groups of colobus and redtail monkeys have been observed to "mob" and chase chimpanzees. Among baboons protective behavior is known to be varied.

When the troop is together, or when 2 or 3 adults form a cluster, both male and female baboons are likely to retaliate initially with vocal and gestural threats. If the predators persist, then brief scuffles may erupt between males of both species. Baboons may lunge at and strike chimpanzees, and they may even grapple briefly at close quarters. However, the few seemingly serious conflicts observed have resulted in no injuries.

If the adult baboons shielding the prey remain silent during pursuit, other baboons are unlikely to become involved even when the activity is visible. The tendency of Gombe baboons to disperse widely when traveling and foraging is probably an added asset to chimpanzees who are sometimes able to continue pursuit for many minutes without interference from other members of the prey community. On the other hand, the threat and/or distress vocalizations of baboons being approached by chimpanzees usually elicit protective responses from other adult baboons in the vicinity. In 7 attempts (54%) that occurred during 1968–69, 2 or more adult baboons joined the prey's cluster to assist, but only after an individual of that cluster had started

8. Refer to second episode on April 27, 1968 in previous chapter.

to *roar, grunt,* or *screech* at the approaching chimpanzees.[9] This convergence upon the scene of pursuit constitutes the main form of collective, or communal, retaliation seen among Gombe baboons. As such, this form is in part analogous to the protective behavior reported for baboons living in the savanna regions of East Africa (DeVore and Hall, 1965). In this connection, a particularly interesting incident was observed at Gombe in 1968 when an adult female serval was mortally wounded by Camp Troop baboons. Suspected of attempting predation upon a young baboon, the serval was severely attacked and lacerated in the hindquarters by many baboons early one evening in lower Kakombe Valley. Chimpanzees at the feeding area at first showed only mild interest in the excitement across the valley, but eventually several crossed the valley to the spot where the cat had withdrawn into a clump of grass. The baboons were already departing when the chimpanzees arrived, and Goliath slapped and stamped upon the still hissing and partly mobile serval. Soon after-wards several female chimpanzees cautiously approached and inspected the cat. These and other chimpanzees waited in the vicinity until dusk, then departed up the ridge. The serval was neither killed nor eaten by the chimpanzees, but died later that evening. This incident serves to illustrate the fact that Gombe baboons—and in this case the same troop that was often preyed upon by chimpanzees—could retaliate swiftly and effectively when confronted by a carnivore. In cases of predation by chimpanzees, it is possible that the absence of harmful biting and slashing by protective baboons was related to the consistent "friendly" interactions observed between members of these species.

The number of baboons participating in protection does not appear to directly affect failure in predation. A single chimpanzee has been observed to seize a baboon in the midst of a troop. But a sudden convergence of many baboons, with a concomitant increase of random activity at close quarters, can effectively divert chimpanzee attention, especially if there are a few direct conflicts requiring defensive counteraction by the chimpanzees themselves. The confusion introduced by the sudden convergence of many agitated, highly active baboons constitutes one aspect of communal retaliation. Another aspect is direct intervention and active physical harassment that leads to momentary distraction of the predators. In situations where the adversities of

9. The two-phase *bark* commonly used by savanna baboons to warn the troop of external dangers such as human beings or carnivores (Hall and DeVore, 1965) was not heard in connection with chimpanzee predation.

environment and disorientation are combined, failure to capture a baboon is common. In 8 cases (61%) observed in 1968–69, the prey individual or the adult in charge of the infant departed unobtrusively while the chimpanzees stood in the midst of milling baboons and looked about confusedly.

It is possible to compare the deterrent of convergence behavior observed at Gombe with the deterrent of dispersal characteristic of some savanna ungulates when carnivores charge. Herd antelopes such as gazelles and hartebeests scatter randomly at high speed when large predators such as lions and cheetahs rush them. With numerous antelopes running and leaping in many directions simultaneously, a predator presumably has difficulty in tracing the movements of a targeted individual. Disorientation at the wrong moment can be crucial to both the chimpanzee predator and the plains predator.

Three particular attempts illustrate and expand some of the above points. On May 14 the prolonged stalk led by Figan progresses smoothly—that is, without excessively disturbing the adult baboons accompanying the prey infant and without alerting other baboons—until his sister Fifi enters the scene suddenly. Fifi's demeanor and movements suggest that she is unaware of Figan's intentions. Whatever the impressions of observers may have been, it is clear that Figan's slow maneuvering to put the prey between himself and Mike fails because Fifi approaches at the wrong moment. These events illustrate several of the points made previously, but it is also the only case on record where one chimpanzee definitely disrupts pursuit by another.

The episode of May 19 provides another example of failure, one which is perhaps more common than the above but has not been closely observed. At 11:05 A.M., with the entire Camp Troop and many chimpanzees present at the feeding area, Figan suddenly walks toward an adult male baboon, Ringo, who carries a *chirp-clicking* black infant, Lamb. Hugo immediately runs to join Figan; Mike advances a few paces some distance across the clearing, then stops and watches. Ringo baboon retreats a few steps from Figan and Hugo, then sits and vocally threatens them from 2 or 3 yards distance. Ringo steps toward the chimpanzees and threatens again more vigorously while holding the infant in one hand; Hugo retreats slightly, and Ringo turns and runs a short distance into the higher grass at the perimeter of the feeding area, where several other male baboons join him. Figan and Hugo follow slowly into the grass but stop some yards short and watch Ringo, who has resumed manipulating and grooming the persistently vocalizing infant. Back in the clearing Mike and other chimpanzees resume other activities. Ringo soon moves farther into the grass and is nearly obscured. Figan, Hugo, and several baboons follow quietly, the male baboons staying between Ringo and the chimpanzees. Moments later sound and action erupt as Stubtail baboon lunges at Figan, and Ringo *roars* and retreats with infant Lamb. Grinner baboon turns suddenly and lunges at Figan, who avoids him. But the activity subsides rapidly, and Figan and Hugo again approach the baboons and sit down quietly about 3 yards from Ringo and the infant. Ringo moves off again at 11:11, taking Lamb ventral; he *roars* once as he looks back at Figan and Hugo, who watch intently for a few moments and then start to self-groom. Ringo settles down in the grass after walking only a few paces and releases Lamb, who finally stops vocalizing as he crawls about on the ground. Soon Figan stands and steps toward them, seemingly ignoring the other baboons nearby; Ringo grabs Lamb and lunges, and Figan *screams* as he backs hurriedly away toward Hugo. The chimpanzees sit again, watching Ringo or the infant; the other male baboons join Ringo in a tight cluster, some watching the chimpanzees. At 11:16 several baboons suddenly advance upon Figan and Hugo, and Stubtail baboon charges Figan with canines bared. Figan makes a silent *fear-grin* and reaches one hand back to Hugo, who touches it to reassure him; Figan is quickly calmed, and Hugo sits down close with a hand resting on Figan's thigh. But Ringo baboon suddenly gets up and lunges at them again, *roaring* and eyelidding; Figan *screams* and hastily steps backwards with Hugo. The male baboons again close in upon Ringo and the infant, and the cluster moves slowly away. Only a few yards farther all the baboons suddenly break into a run downslope toward Palm Grove. Figan and Hugo immediately pursue, and other chimpanzees descend from trees and follow. All disappear silently into the tall grass. Observers hurry downslope, and at 11:20 A.M. they overtake the chimpanzees, who are sitting quietly at the edge of Palm Grove, looking into the forest. The baboons are nowhere in sight, and the pursuit has clearly ended.

Two aspects of this attempt episode are notable. First, no chimpanzee other than Hugo—whose participation seemed less than wholehearted—became actively interested during Figan's 15-minute attempt, despite the fact that many—including several adult males—were present at the feeding area and aware that pursuit was in progress. General disinterest such as this is uncommon, so it is possible that other chimpanzees anticipated failure from the very beginning! Second, Figan persisted in his actions toward the infant baboon despite his being repeatedly intimidated by the threats of Ringo and other baboons. Not only was such extensive and prolonged protective behavior rare in other episodes at Gombe, but it is even more exceptional that a cluster of male baboons, who are capable of highly effective retaliation upon a serval cat and who as a species have a reputation for efficient collective retaliation against far more formidable predators on the open plains, do not attack and injure chimpanzees attempting to capture an infant baboon.

A third interesting episode occurred on January 14, 1969. Late in the morning of this nonfeeding day, 3 adult male chimpanzees groom quietly in a session. Only 3 other chimpanzees are present, but baboons are dispersed throughout the feeding area, resting or foraging. One of these, a mother with a transitional infant, wanders about only a few yards from Mike, Leakey, and Hugo. These chimpanzees occasionally pause during grooming to glance up at the baboon mother and infant, and also show mild interest in infant baboon vocalizations farther away in the brush surrounding the feeding area. Some minutes later the troop begins to drift away from the feeding area, one by one entering the high grass at the perimeter. Hugo watches several leave, then suddenly rises, and runs silently after the female with other chimpanzees following. Several baboons immediately begin to *roar* and *bark* as they stop to face the approaching chimpanzees, who stop suddenly about 5 yards short of the mother and infant. One adult male baboon charges Mike and succeeds in chasing him some distance across the clearing; another threatens and chases Hugo onto the roof of a building, where Hugo turns and agitatedly threatens back while *screaming* loudly; and 2 more baboons lunge at Leakey, striking his back. The attempt is clearly over. The adversaries watch each other intently for a few minutes, then the baboons depart from the feeding area.

This brief episode illustrates how convergence behavior can deflect predatory intent. The attempt was exceptional in that baboons succeeded in actually routing the predators without inflicting wounds.

In summary, predatory failure may be associated with numerous variables: (a) loss of interest on the part of the predators; (b) interference of adverse environmental conditions; (c) evasion by the prey; and (d) retaliation by the prey or the prey's community. These factors usually operate in combinations, and their disruptive capacities are sometimes supplemented by other special conditions.

INDICATORS OF INTEREST AND INTENT

The signs of interest and intent shown by chimpanzees before pursuit begins often serve as cues to other nearby chimpanzees as well as observers. Chimpanzees receptive to these cues become instantly more alert and excited, and awareness is at times so acute that individuals seem to start pursuit in unison even when widely dispersed. For humans, quick and correct recognition of predatory intent depends upon an observer's sensitivity to the surrounding conditions, as well as to differing characteristics and varying moods in individual chimpanzees. After an observer becomes thoroughly familiar with these idiosyncrasies, the subtle signs and actions expressing predatory intent can be recognized as readily as in human interactions.

Generally speaking, predatory interest and intent among chimpanzees are shown by a concentratedly set or often blank facial expression, a prolonged stare fixed upon or directed toward the target prey, and a tensed posture with partial hair erection all over the body. These signs, which are followed by the rapid and very fluid locomotion of pursuit, differ subtly from comparable elements common to other social activity in that expressions are highly attentive and somewhat tensed and movements appear exceptionally deliberate, directional, and stealthy. For example, the fixed stare of predatory interest is usually distinguishable, given adequate allowance for individual variability, from the fixed glare of mildly agonistic social interaction described by Lawick-Goodall (1968b). However, such distinctions are not always clear in the field situation because they tend to vary by degree, and because individual variation in physiognomy and expression may blur the subtle signs that can appear in the setting of a jaw or the position of a lower lip, the raising or lowering of an eyebrow, the extent of hair erection, and the tensing of limbs in anticipation of action (Plate 21). Vocalizations and gestures are of negligible assistance in recognizing predatory intent because neither appear to be stylized for this behavior. Barely audible single-phase *hoos* are sometimes given by adult chimpanzees just prior to starting pursuit, but these calls are heard in a variety of other circumstances as well.

PLATE 21A

Mike's posture and expression are indicative of interest in the activities of other chimpanzees: here the approach of new arrivals at the feeding area. Note the extent of hair erection (including the head), the partially slack face with lower lip drooping, the relaxed body posture in the first frame, and the fixed glare of slightly aggressive expectancy in the second frame.

PLATE 21B

Hugh's posture and expression are typical indicators of interest in predation, and are often the first signs of a pending chase. Note here the expectant forward lean of the body, the partial hair erection, the blank face and fixed stare with mouth closed. Then compare to similar components in Mike's appearance.

STAGE 1: PURSUIT

The activities of the first stage of predation can be divided into modes of seizing, chasing, and stalking prey. Among the 30 episodes observed in 1968–69, the modes of only 22 could be classified with some accuracy (fig. 5). The activities of pursuit are, of course, part of a continuum even though the modes appear to be discrete in many episodes: stalking and chasing can end with seizure, and stalking can grade into chasing.

Seizures are somewhat unique because the initial signs of interest and intent, as well as the actions of pursuit, are greatly condensed and may not even occur. In general terms, seizure involves explosive action by the predator (s) in taking full advantage of exceptionally promising circumstances that arise suddenly. Mode–1 pursuit thus requires little expenditure of energy by the predator and minimal control over circumstances. But seizure is usually dependent upon special situations: where predator and prey are intermingling freely in close proximity, where the prey individual and other members of its community are inattentive or distracted, and where visual predator-

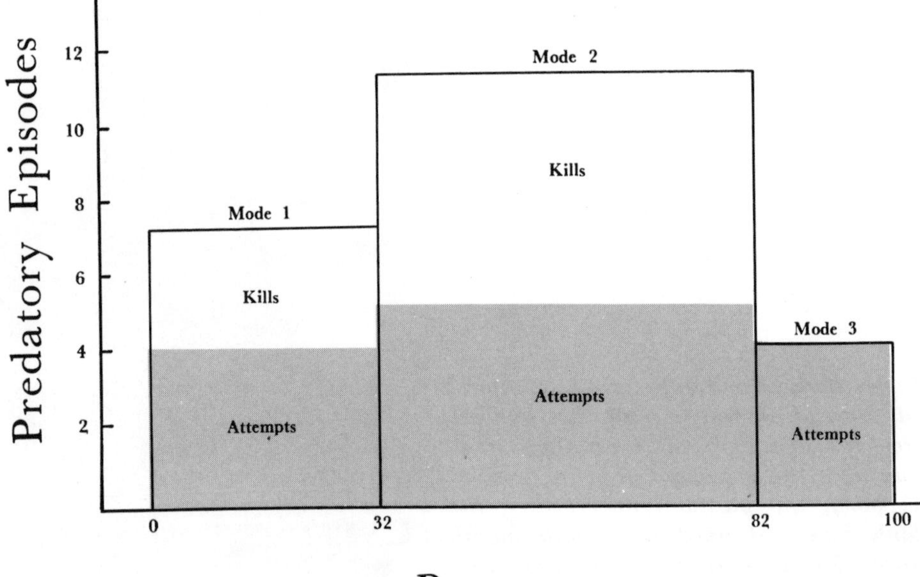

Figure 5. The frequency of predatory success in terms of pursuit modes, where 100% includes 22 episodes observed during 1968-69.

prey contact can be easily maintained. In some respects this "opportunistic" snatching of prey seems more a matter of food collection than true predation.

Mode–2 and mode–3 pursuit differ from mode–1 in that the actions of the predator (s) are more deliberate and controlled. Two components, duration and distance, become important. The extremely rapid, explosive quality of seizures becomes, at the other end of the continuum, a matter of prolonged strategy and maneuvering. Similarly, the initial distance between predator and prey at the start of pursuit increases from a few yards maximum in seizures to a maximum of several hundred yards in chasing or stalking. An important behavioral component of mode–2 and mode–3 is that several chimpanzees are likely to coordinate their actions for a single purpose. In the stalking of prey, this coordination often tends to approach active cooperation, where the movements of various participants are deliberately complementary.

The pursuit—and the capture—of prey is primarily the domain of mature male chimpanzees. With males present, no female was observed participating in predation prior to the stage of consumption during the 1968–69 period. However, females have been observed to pursue and capture prey on some occasions. In the late afternoon of April 27, 1970, for example, Athena and Winkle captured 2 piglets at the base of a ravine (KK 1 on Map IV) in Kakombe Valley. Only one other chimpanzee, also a female, joined the meat-eating cluster that day, so male responses to such a situation could not be observed. It is possible that females pursue prey only in the absence of mature males.

A significant element of all modes of pursuit is silence on the part of the predator (s) until seizure is attempted. As complete silence is very rare in other situations involving many active chimpanzees, the absence of noise is a reliable indicator of predation.

a. *Seizing Prey*

Because of the explosive nature of this mode, pursuit is often completed before the situation registers with observers. However, a general picture of seizure can be reconstructed from 7 pursuits (32% of 22) observed in detail during 1968–69,[10] of which 6 were documented at or near the feeding area (Appendix A, Map V) probably because observers were consistently present in this area. Seizures were 4 times

10. See Appendix B, Table III for a listing of modes.

associated with banana distribution, after which large numbers of chimpanzees and baboons were in close proximity; in 2 cases seizure occurred in a similar situation but without bananas; and in 1 case there were neither bananas nor a large number of primates.

When Mike captured the baboon infant Amber on March 19, 1968, tension was high among expectant chimpanzees before the feeding boxes were opened, and Mike's frustration in not getting enough fruit may have influenced his actions. In the other episodes, seizure occurred when the chimpanzees were relaxed and seemingly self-occupied. On April 15, for example, Mike and Hugo are grooming in a mutual session near the center of the feeding area, more than 2 hours after eating bananas, when an infant baboon, Lane, starts a tantrum nearby. The rest of Beach Troop is dispersed throughout the clearing, many searching in the grass for discarded bananas and peels. An adult male baboon, Moon, appears to be the cause of Lane's *gecks* and *screeches*. Perhaps becoming irritated by the tantrum, Moon cuffs and tumbles Lane repeatedly, whereupon Lane intensifies his cries. Mike and Hugo interrupt grooming several times and look briefly at Lane, and once they stand as though about to advance toward the baboons; but each time grooming is resumed, and the baboons apparently ignored. Both Moon and Lane appear to be so occupied with one another that neither pays attention to the nearby chimpanzees. After 12 minutes of sporadic tantrums from Lane, 4 adult male chimpanzees—Mike, Hugo, Charlie, and Rix—suddenly rush toward the infant. Moon instantly takes Lane ventral, and baboons scatter in all directions. Several chimpanzees watching from a distance *pant-hoot* and *waa*. The male chimpanzees stop, all standing with hair on end, and watch the baboons disappear from the feeding area. Some moments later the same males rush into the high grass, and Hugo eventually emerges with a different baboon infant, Beau, in one hand. Consumption of Beau begins soon afterward.

A second episode, observed on June 19, is interesting from the standpoint that a baboon mother persistently ignores the danger of staying close to agitated chimpanzees. With banana distribution still in progress at 10:42 A.M., Mike and Hugo begin to arm-threaten and then slap an adult male baboon, Mandrill, who has been taking peels from the ground nearby. A female baboon, Sif, also jumps away from the threats but stops barely 5 yards away; her 8-week-old infant, Thor, is ventral. Mandrill soon walks back to the chimpanzees and resumes collecting peels, and Sif follows and watches. Holding Thor to her abdomen, Sif stands less than a yard away from Mike, who seems to ignore them; both the mother and infant are silent. Suddenly Mandrill leaps back again as Mike rises bipedally to threaten him, and Sif also retreats a few steps. Mandrill returns almost immediately to the chimpanzees and searches for more peels. Two other chimpanzees, Charlie and Goliath, finish their bananas at this point and lie down a few yards away. Mike and Hugo also finish, and both begin to groom Hugh. Two more adult male baboons and several juveniles join Mandrill and Sif, so the mixed group (all are within a circle of 10 yards' diameter) now includes 5 mature male chimpanzees and 7 baboons of various ages. Thor is the only infant present. All seem quite relaxed: several chimpanzees groom mutually, and Salty baboon begins to groom Sif. Mike suddenly *waas* and arm-threatens Mandrill again at 11:02; Mandrill wheels and returns the threat by eyelidding at Mike, who in turn slaps the baboon's nose; Mandrill leaps backward, and calms down rapidly. With Thor ventral and in full view, Sif walks to within approximately 2 yards of the chimpanzees as she collects banana scraps; she gradually passes near Mike, Hugh, and Hugo, who seem to ignore her, and eventually settles down to groom Mandrill. At 11:07 Mandrill takes infant Thor from his mother, who does not pause in grooming the adult male. The baboons are only a yard away from the chimpanzees. Then at 11:09, several male chimpanzees—Mike, Hugh, Hugo, and Charlie—suddenly pounce on Mandrill, grab Thor from his hands, and begin immediately to tear the infant apart as they huddle in a tight cluster. The baboons—including mother Sif—scatter quickly. The only baboon who stays near the chimpanzees is Mandrill, who *barks* repeatedly as he pushes with both hands against one chimpanzee's hunched back, but to no apparent avail. Thor is soon halved by Mike and Hugh, and other chimpanzees follow as these two walk away from the scene, climb nearby trees, and begin to consume the infant.

Both of these seizures occurred on feeding days, but only one while bananas were still being distributed. In both cases numerous baboons and chimpanzees were present: all 50–55 members of Beach Troop and 28 chimpanzees on April 15, and 70–75 members of Camp Troop and 34 chimpanzees on June 19. Vocalizing infant baboons, each only 1 or 2 yards away from the chimpanzees, were preyed upon both times, once by male chimpanzees who appeared quite relaxed and once by males who were irritated by the persistent approach of baboons wanting banana scraps. A considerable span of time elapsed in both cases between the initial availability of a young baboon and the moment of seizure, but the act of seizure was explosive.

Seizures can also occur in fortuitous conditions, as, for example, when individuals meet suddenly in heavy foliage. This was the case on October 4 when Faben seized Lane baboon in Palm Grove. Because of great diversity in setting, seizures can only be generally characterized as explosive responses to very favorable situations which yield accessible prey. If the initial lunge or dash is unsuccessful, then pursuit is likely to be abandoned and other activities resumed.

b. *Chasing Prey*

Of the 22 episodes in which pursuit was observed in 1968–69, 50% can be classed as mode–2. However, only 8 of these 11 cases are used to sketch a general picture of chasing because 3 cases were poorly documented.

One excellent example of the mode–2 pursuit—the episode on December 23, 1968—has been presented in detail in the previous chapter. An even more spectacular chase was observed on the morning of May 19, 1968. This episode begins at 7:12 A.M., when an adult male baboon, Ringo, sits down about 3 yards away from Charlie, who leisurely eats bananas. Ringo holds ventral a black infant that soon begins to *chirp-click* as it struggles to get away from him. Figan watches them briefly from halfway across the feeding area, then starts to advance. Suddenly Charlie lunges toward Ringo, who runs off across the clearing with Charlie close on his heels. Figan also starts to run, but he soon drops behind and stops to watch the chase. Moving at high speed, Ringo and Charlie cross the clearing, turn down the other side, and return across the lower slope.[11]

As they make the second turn, 4 or 5 male baboons burst from nearby bushes and intercept Charlie, who raises both arms to head level and slaps baboons aside without breaking stride. Emerging from the cluster in a fraction of a second, Charlie continues after Ringo baboon, who has just scrambled into the branches of a low tree. Threatening and *roaring,* the other baboons wheel in pursuit of Charlie, who now leaps, with hands outstretched in a long dive for Ringo's branch. While Charlie is still airborne, Ringo jumps down the other side of the tree. Charlie responds instantly by using the branch as a pivot: he grasps it with both hands, swings around it, releases his grip on the way up, flips through the air in a full gainer

11. Both Ringo and Charlie covered 300 feet in about 7 seconds—a speed of nearly 30 miles per hour. Estimated maximum speeds for other predators are 30 mph for the lion, 40 mph for the wild dog, and 55 mph for the cheetah (Kruuk and Turner, 1967), and 40 mph for the hyena (Kruuk, 1966).

on the way down, lands on both feet beneath the tree, and continues the chase. He immediately disappears after Ringo into the high grass surrounding the feeding area. But Charlie is overtaken by the other baboons; he turns to face them, *waas* a few times, and slaps several baboons. After a brief interval of milling about in the grass, everyone begins to calm down. At 7:18 Charlie returns and displays across the clearing, dispersing several chimpanzees along the way, and disappears into the grass at the opposite perimeter. Ringo baboon must in the meantime have circled around in the grass because he now emerges near Charlie, still holding the infant ventral. Charlie leaps high into the air with both arms flailing, *waas* several times, and again starts to pursue Ringo. A second male baboon intercepts and attacks Charlie, lunging and slapping at him. Rix immediately joins the fight, and they push and strike at one another for several seconds. No longer chased, Ringo stops about 10 yards away, sits down, and watches the conflict, and the infant resumes vocalizing and struggling. When the fight breaks up, Charlie sits down in the grass. Ringo walks back toward him and calmly settles down a few yards away. Charlie ignores them this time and soon returns to his bananas.

This episode illustrates the tenacity and persistence of chases. Both the time that elapsed and the distance that was covered, as well as the energy expended by the predator, was greater than in mode–1 pursuit. The effective interference of convergence by the prey's community is also demonstrated. However, the utility of this form of protection has been reduced in other chases by collective pursuit by chimpanzees.

c. ***Stalking Prey***

Two samples of mode–3 pursuit—the episodes of April 27 and May 14, 1968—have been presented in the previous chapter. However, the clearest example of this mode is an episode observed some years ago by Lawick-Goodall (1968b: 191), when Figan, then an adolescent male, stalks a juvenile baboon. As Figan slowly climbs a palm tree containing the lone baboon, other male chimpanzees resting and grooming nearby stand and approach the tree. Some move to the base of the tree, and others disperse to adjacent trees "which could have provided an escape route for the quarry." Probably alarmed by Figan's stealthy approach, the baboon climbs to an adjacent palm tree and then, when Figan begins to follow slowly, returns to the first tree and starts to vocalize. Approached again, the baboon leaps to a smaller tree nearby. Another chimpanzee, who had been quietly waiting beneath this tree, immediately picks up the pursuit by climbing quickly toward the baboon. The other chimpanzees begin to adjust when the baboon suddenly leaps some 20 feet to the ground. Though chased a short distance by those on the ground, the juvenile baboon escapes and rejoins the troop nearby.

This mode of pursuit differs from chases mainly in degree. Stalking is usually longer in duration than other modes, and appears more deliberate and more cooperative. Lacking the explosive, high-speed actions of the other modes, stalking entails a stealthy, disciplined intent to corner prey without alerting it to danger. The average duration of mode–3 pursuit was 28 minutes in episodes recorded during 1968–69.

Despite complete silence on the part of the predators, these stalking movements often progress with considerable coordination among the participants. The cluster appears to move as a fluid unit with each chimpanzee positioning and repositioning himself as necessary to maintain the enclosure by anticipating escape routes. Direct harassment of the prey appears to be controlled by a single chimpanzee who operates within the enclosure, but control may pass through several individuals in a short period as required by the constantly changing conditions of pursuit. Using the escape movements of the prey as cues, *control individuals* can replace one another quickly and silently, minimizing interference within the stalking cluster. That the top-ranking male (Mike) of the study community is not always the control individual appears to be an important aspect of these tactics because competition for the best position during pursuit would presumably reduce efficiency.

Stalking was observed only 4 times (18% of 22) during 1968–69 and has been relatively infrequently observed in other years as well. During that year, all episodes of this type were failures. In fact, the success rate for stalking (00%) was considerably lower than were the rates for seizing (43%) and for chasing (55%) prey. A problem arises here of learning why stalking is practiced if it is likely to fail much more often than other modes of pursuit. Although this must remain an open question at this stage of research, the very fact that chimpanzees practice stalking against considerable odds would seem to emphasize their motivation for predation.

STAGE 2: CAPTURE

The climax of pursuit is the capture of prey—the climactic moment when the level of activity and vocalization rises sharply among chimpanzees. When a large number of chimpanzees is present in the vicinity of capture, this outburst of excitement tends to strongly arouse observers as well. Unfortunately, however, the details of capture are rarely observed in comparison with pursuit and consumption for a variety of reasons: some captures are in distant places and, by the time an observer arrives at the scene, consumption is well underway; others happen unexpectedly or very rapidly, or are obscured by foliage, and are consequently missed even at close range; and still others are not seen because observers are easily outdistanced during pursuit. Thus, although 132 cases of predation have been recorded at Gombe, predatory *behavior* was observed only 83 times, and the particular activities of capture were seen on about 12 occasions. During the 1968–69 period, only 4 captures out of a total of 12 were observed in detail.[12] The low frequency of observation is also related to the short duration of captures, which average less than 4 minutes when measured from the moment prey are acquired to the moment carcasses have been divided into major portions.

Despite its brevity, the second stage of predation includes several sequential components: the acquiring, killing, and dividing of individual prey. Diversity within the sample is wide, both in situation and in behavior, so little can be concluded about patterns and preferences.

Two captures from early years have been reported by Lawick-Goodall (1968b)—one made by an adult male chimpanzee who grabbed a juvenile baboon by the leg and battered its head against the ground while standing bipedally; and another by an adolescent male who climbed a tree toward a troop of colobus monkeys, ran along a branch to snatch one of the monkeys, and presumably broke its neck or choked it to death. On March 19, 1968, Mike captured an infant baboon by approaching the mother from above and slightly behind and snatching it from her abdomen. Although the infant was killed later by another chimpanzee, Mike did slam the baboon against the ground several times as he ran bipedally across the clearing. On April 29, 1968, Charlie captured an infant carried by an adult male baboon that was cornered in the crown of a palm tree. Charlie approached the adult baboon very cautiously, walking bipedally along

12. March 19, April 29, June 19, and October 4, 1968

a palm frond while holding others for support; the baboon backed as far as possible along the frond before Charlie suddenly leaned forward to snatch the infant away with one hand; still holding the live captive, he slid rapidly down a frond and disappeared from sight. A burst of chimpanzee and baboon sounds followed, together with much activity, but observers were delayed while crossing the valley and the baboon was already dead when they arrived. Another capture, on June 19, 1968, involved 4 adult male chimpanzees who took an infant from an adult male baboon. The capture occurred at the feeding area when many other baboons were nearby, and the captive was torn apart immediately as the chimpanzees tried to obtain portions. Still another capture was observed on October 4, 1968, when Faben snatched a juvenile baboon in Palm Grove and displayed bipedally along the forest floor while swinging the captive overhead by one leg and smashing the head repeatedly against tree trunks. The baboon was killed some minutes later by Mike, who bit the back of its neck the moment Faben relinquished it.

a. *Acquiring Prey*

Once the distance between predator and prey has been reduced to a yard or less, chimpanzees are inclined to lunge at the prey suddenly, using one or both hands to make the capture. At the moment of acquisition the predator's attention appears to be focused entirely upon the prey individual even when it is accompanied by adults. Attacks upon the adults accompanying the targeted prey individual have not been observed. Because misses are rare at such close quarters, the prime concern of predator chimpanzees must simply be to get close enough to the selected individual for a sudden lunge or grab. Location appears to play little or no part in acquisition, as the settings varied from open grass to dense thickets and from broken terrain to the tops of trees.

b. *Killing Prey*

Captives tend to be killed within seconds of acquisition, but there can be extenuating circumstances. On March 19, for example, a delay was caused by the convergence of numerous chimpanzees and baboons, and on April 15 consumption was underway for more than half an hour by the time the captive presumably bled to death. The techniques used to kill prey vary: (a) when a single chimpanzee obtains the captive, killing may take the form of a quick biting or snapping of the neck; (b) a single chimpanzee may also kill by beating the captive's head against a solid surface, such as the ground or tree trunks, while flailing it by the legs; (c) a captive that is acquired simultaneously, or nearly so, by several predators may literally be torn apart as each chimpanzee tugs on the body or limbs to obtain portions.

The Gombe chimpanzees have been observed to use only hands and teeth in killing prey. Unlike the attack responses recorded by Kortlandt (1963, 1965, 1967) in field experiments displaying stuffed leopards to chimpanzees, no objects of any kind—neither missiles nor clubs—were picked up or employed purposively in predatory episodes at Gombe. However, the same chimpanzees who practice predation at Gombe use such objects as grass stalks, vines, sticks, and leaves in a variety of activities that include insect-eating, drinking, wiping the body, and investigative probing; and such objects as stones, branches and palm fronds are frequently used during displays (Lawick-Goodall, 1965b, 1968b).

The relatively meager data on killing behavior permit only estimates about performance patterns. There is some possibility that individual habits, not just circumstance, influence the method used on any given occasion. For example, in episodes where Mike was a participant in captures, the prey appeared to be killed more rapidly and efficiently than when other chimpanzees, such as Hugo, obtained the prey because Mike is likely to bite or break the neck almost as soon as he acquires the prey.[13] Although one can only infer that some chimpanzees have developed preferences or idiosyncratic techniques for killing prey, this hypothesis is in line with the highly individualistic behavior exhibited by Gombe chimpanzees in many other maintenance and social activities.

c. *Dividing Carcasses*

The discovery that freshly killed prey are often distributed, or shared, among many individuals was first made by Goodall (1963a). However, the topic of this section is carcass division, not general meat distribution. As will be seen in the subsequent discussion of consumption, the terms *distribution* and *division* are not synonymous because no transactions occur between individuals in the latter case. In other words, the activities during capture are more a matter of

13. On April 15 the prey slowly bled to death in Hugo's hands, and on October 4 the prey was not killed until Mike obtained it from Faben.

appropriation, with individuals tearing portions from the carcass.

Of the 12 prey killed in 1968–69, 8 were divided by several mature males after collective acquisition, 2 were obtained by a single male who later shared some of the meat with others, and 2 cases remained uncertain.[14] In all cases of division, carcasses were split into large portions such as forequarters and hindquarters, or into arms, legs, and torso. Most division occurred within 3 to 4 minutes of acquiring the prey, and with 2 to 4 chimpanzees getting major portions each time. The timing of division appears to be a critical factor, because a chimpanzee apparently must participate in acquiring or killing the prey, or at least be able to get his hands on the prey immediately afterwards, in order to be eligible for obtaining a large portion without rebuff from others holding the carcass. If a single chimpanzee succeeds in acquiring the prey—by hoarding it or moving away from the others rapidly—then the other chimpanzees are unlikely to attempt to take it away or to appropriate a large portion by division. If, on the other hand, several manage to grasp the prey while it is still "common property," then each can pull simultaneously, and apparently without risking objection from any of the other individuals. This does not imply, of course, that all participants will inevitably get large portions, for carcasses tear apart in a variety of ways. But no single chimpanzee is likely to assert himself in aggressive competition, regardless of social rank, by physically threatening or attacking another individual who also has his hands on the prey. A considerable amount of energy and effort may be expended in tearing apart carcasses, but no overt conflicts among members of such clusters were observed during 1968–69, and very few in other years.

Once a carcass is ripped apart and divided among several individuals in the above manner, the cluster usually disperses as each participant in possession of some portion moves slightly away to begin consumption. Those chimpanzees who still lack meat at this point tend—again regardless of relative social rank—to respect proprietary rights and to *request* meat from others. Usually this is the time, when portioning has stabilized, that other chimpanzees who have meanwhile converged upon the capture site—mostly females and subadults—begin to form food-sharing clusters around these central figures. An exception to this rule on many occasions was Flo, the oldest and highest ranking female with the largest family, who regularly joined the adult male clusters during division of carcasses.

---

14. The 2 cases with solitary chimpanzees: April 15 (Hugo) and June 30 (Goliath). The 2 uncertain cases: March 19 and December 3. Refer to diagrams in Appendix C for details.

STAGE 3: CONSUMPTION

The final stage of predation may begin immediately after the kill if one chimpanzee has the prey, or after the initial division of the carcass if several pull it apart. Only when the entire carcass has been consumed, or when all the chimpanzees lose interest in meat, does the stage end completely. In fact, loss of interest probably never occurs because chimpanzees who are at first unable to obtain meat will retrieve all the remains discarded by others.

Out of the 47 hours of predatory behavior recorded in the 1968–69 period, 43 hours were spent observing consumption. Being the longest stage of predation by far, consumption lasts 3.5 hours on the average, with an observed minimum of 1.5 hours and a maximum of nearly 9.0 hours (fig. 6). Variation in the duration of consumption appears to correlate less with prey size, which is fairly uniform in all episodes, than with the total number of individuals who initially obtain large portions and who eventually participate in distribution. Prolonged consumption periods tend to involve the largest number of participants (fig. 7), even though this means that actual consumption for each chimpanzee is lower during long episodes. For example, the

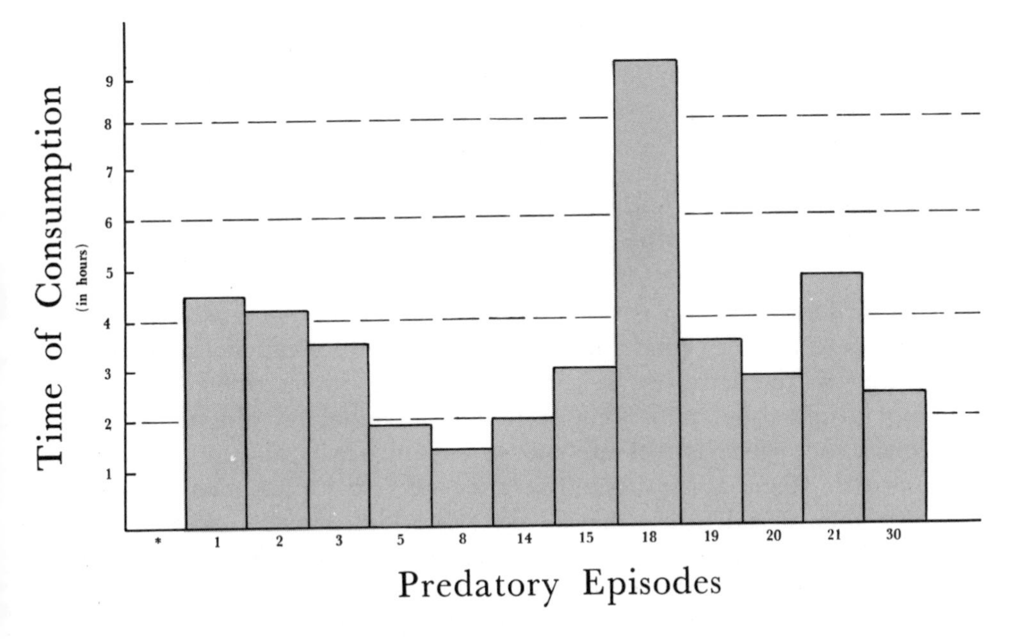

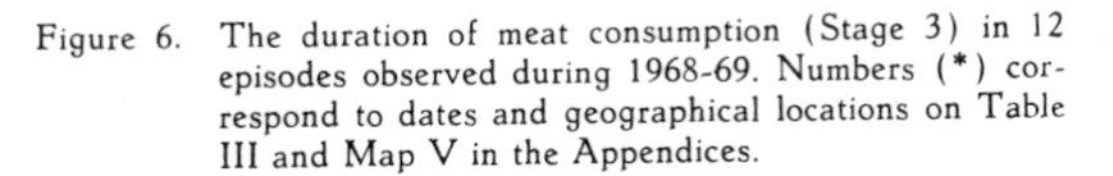
Figure 6. The duration of meat consumption (Stage 3) in 12 episodes observed during 1968-69. Numbers (*) correspond to dates and geographical locations on Table III and Map V in the Appendices.

August 7, 1968 consumption (no. 18 in figs. 6 and 7) lasted 8.8 hours and 15 individuals ate some portion of the colobus monkey, whereas the June 30 consumption (no. 15 in figs. 6 and 7) lasted 2.9 hours and was attended by only 5 individuals. The emphasis is apparently on savoring and sharing meat, not simply on consuming the greatest amount in the shortest time.

Average participation in the consumption periods observed in 1968–69 was 8 chimpanzees per episode, with extremes of 4 and 15 individuals. These figures are much higher than the average number that initially obtained meat from carcass divisions, so it is clear that most chimpanzees obtain meat through distribution during the final stage of predation. Fully 38 chimpanzees, or 75% of the entire study community, participated in consumption at least once during the year—a figure that breaks down nearly uniformly into 37% adult males, 33% adult females, and 30% subadults. Individuals are apparently more subject to discrimination than are age-sex classes. Only 5 adults did not eat meat at any time during the year. Although 3 subadults below the age of 5 years also did not eat meat, other subadults of 2 years of age did obtain some occasionally.

The number of hours spent in consuming meat or in waiting and attempting to obtain it varies greatly among individuals, as illustrated by the large range within the adult male class (fig. 8). Most males attended and participated in consumption fairly consistently, whereas only one female, Flo, who was in general the most privileged among females, approached the high attendance rates of several adult males. These differences are, among other factors, probably a reflection of the great variation known in individual ranging habits, for distance from the kill site would greatly influence attendance in such rugged terrain.

A striking aspect of consumption is the rapidity with which chimpanzees calm down after the excitement of killing and dividing prey, as well as the leisurely, relaxed atmosphere which prevails during meat distribution. After one or more individuals obtain major portions, other chimpanzees of both sexes and nearly all ages begin to approach them, some hesitantly and others with assurance. When more than 2 individuals have sizable portions, clusters of interested chimpanzees soon collect around these and begin to watch, request, or take meat. These meat-eating clusters rarely exceed 3 or 4 individuals at one time, although several clusters may operate simultaneously. In addition there is likely to be a number of nonparticipant chimpanzees nearby—ones who may pass the time in grooming, look-

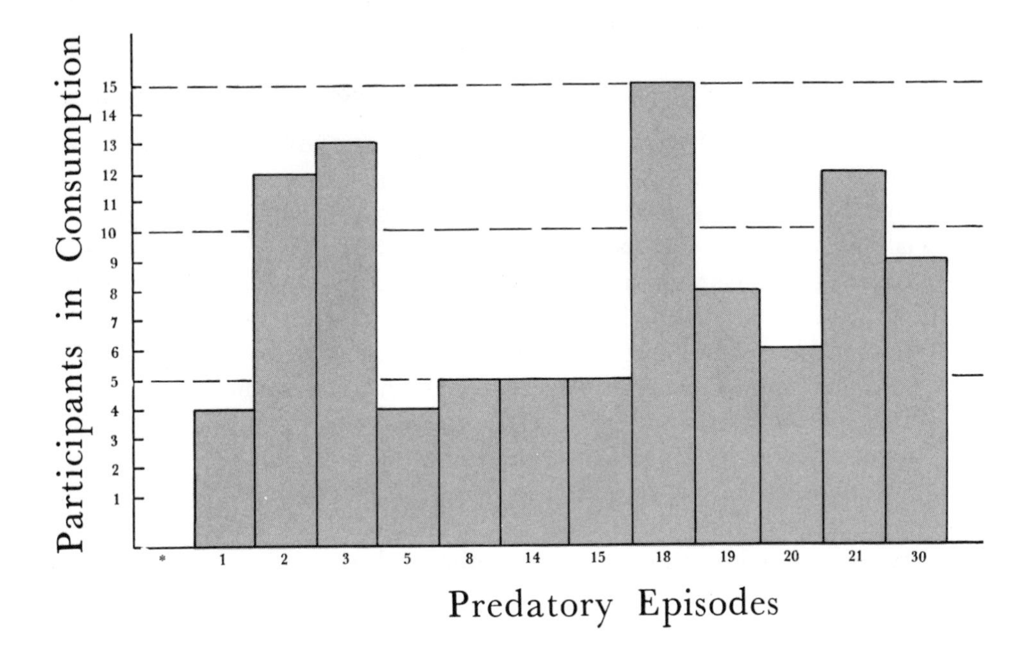

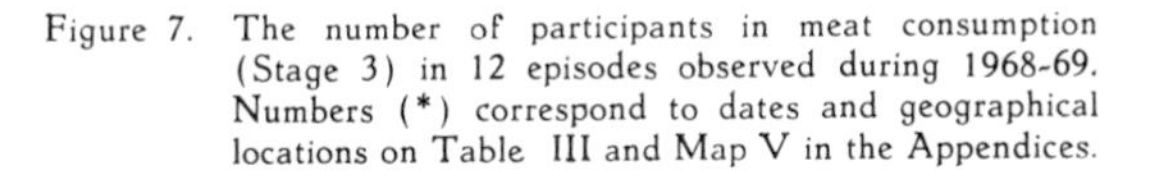
Figure 7. The number of participants in meat consumption (Stage 3) in 12 episodes observed during 1968-69. Numbers (*) correspond to dates and geographical locations on Table III and Map V in the Appendices.

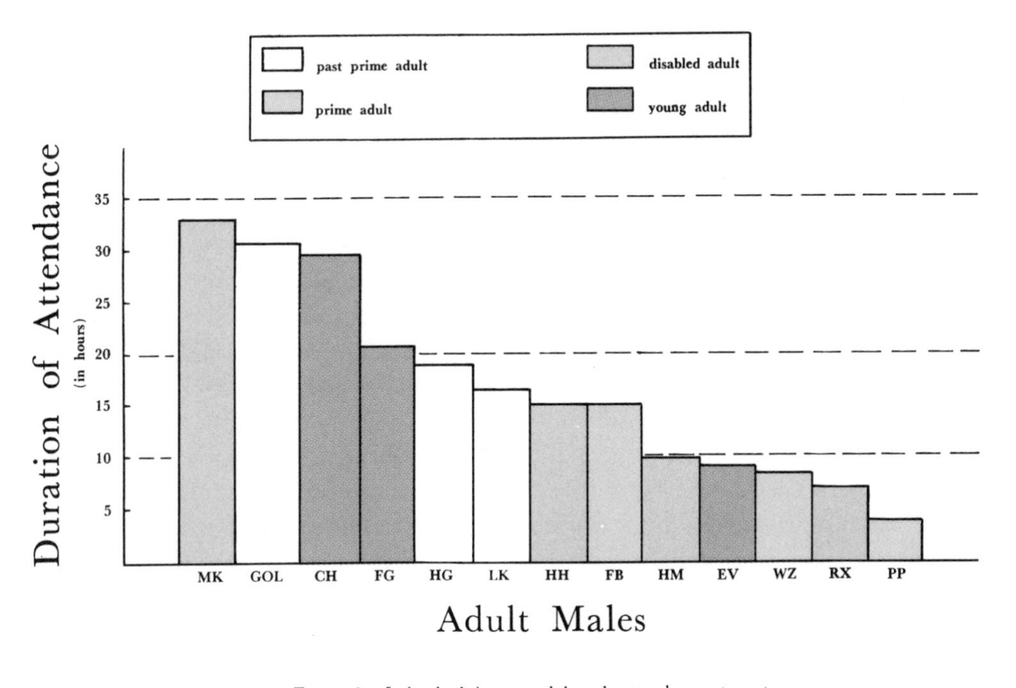

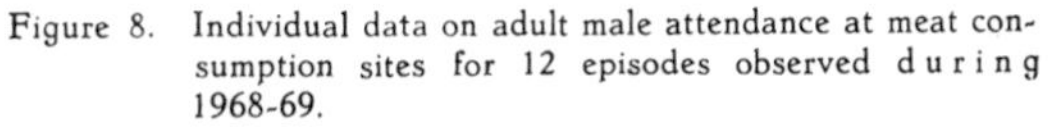
Figure 8. Individual data on adult male attendance at meat consumption sites for 12 episodes observed during 1968-69.

ing on from a distance, wandering about in search of dropped fragments, or just sitting quietly. Clusters tend not to stay constant in composition and size for any length of time, particularly if attendance is high and if several have large portions, because chimpanzees whose requests or advances are unrewarded by one adult are likely to try with another, while still others may join clusters only sporadically. Rather than being random, however, cluster formation appears to be regulated by many subtle factors, such as personality and mood, and special status and role relationships. Thus, a chimpanzee who regularly attempts to obtain meat from one individual (with or without success) often does not even approach another that also has meat, and a chimpanzee who acts confidently in the presence of the individual holding meat may nonetheless leave the cluster if certain others join it.

In many cases, then, consumption involves a continuous shifting and blending of clusters that dissolve and reform around those holding portions of a carcass. Excitement and activity may explode periodically, but an aura of calm and relaxation is most likely to prevail. Frequent change in clusters and in other social activities is possible because individuals that possess portions tend to disperse only slightly from the kill site. Often all of the carcass is consumed within a small area of a few dozen square yards or within a single tree.

a. *Dismantling the Carcass*

Like episode duration, the steps of dismantling and consuming prey vary in accordance with the number that initially obtains large portions and the number that subsequently participates in meat distribution. A single chimpanzee has never been observed to eat an entire carcass. Disregarding the episodes of March 19 and June 19, 1968, when the carcasses could not be regularly observed, the remaining 10 cases observed in 1968–69 show the following general sequence of consumption: the viscera, the chest and ribcage, the appendages, and the head. These steps are unfortunately less than precise, however, because more than one chimpanzee usually obtains portions, and the observations can only be collated with difficulty. Sometimes there appears to be no sequence at all, as on April 8, 1968, when several chimpanzees eat simultaneously from all parts of the carcass for more than half an hour before the remainder is divided. More commonly, the forequarters and hindquarters are separated at the waist during division, with one chimpanzee obtaining each half.[15] In other instances the head, sections of the torso, and the appendages are appropriated separately.[16] Dismantling of a carcass is likely to begin by opening the underbelly and immediately removing all the internal organs. This is probably more a result of many individuals grabbing for meat than a matter of preference, for the viscera were not touched until much later on those occasions when one chimpanzee acquired the entire carcass.[17] Once the internal organs are removed, consumption is likely to focus upon the exposed section of the chest cavity and abdomen. Strips of meat and bone are removed by means of fingers and teeth, and the body skin may be gradually peeled back over the spine and shoulders in the process. Consumption of the appendages may overlap these activities or begin only after most of the torso is finished. The fingers and toes, and eventually the hands, feet, and the tail, are usually eaten before the muscles and bones of the arms and legs. Although the brain may be extracted at any point of the sequence, the head—and especially the jaw and braincase—is commonly the last to be eaten, along with the bits of bone and strips of skin that remain.[18]

Teeth, hands, and sometimes feet are used by chimpanzees to divide a carcass. Normally, unbroken skin is punctured and ripped away with the canines and incisors; exposed flesh may be stripped away with the mouth or hands. Small bones are thoroughly cleaned by sucking and scraping and are then chewed apart or discarded (and collected by others) ; large bones such as those of the arms and legs are cracked between the molars, and the marrow is sucked out while the bones themselves are gradually consumed. All internal organs, connective tissue and cartilage, bones and skin, and often even the nails, teeth, and hair are eventually consumed. Analyzed feces have contained all of these materials. Dropped or discarded fragments are usually collected as soon as possible by chimpanzees who cannot or dare not obtain meat by interacting directly with one that possesses a portion. Subadults in particular will meticulously search all about the site after the adults move away. Consequently, even small pieces are taken away and dispersed beyond recovery. Careful search of sites after chimpanzees departed never yielded more than a few tiny, usually unidentifiable bits of bone or skin.

15. April 29 and December 23, 1968, and March 11, 1969.
16. April 2 and October 4, 1968.
17. April 15 and June 30, 1968.
18. Illustrations appear in the October 4 episode in the preceding chapter.

Aside from the thoroughness of consumption, kill remains would not likely survive the rigors of decay, insects, and scavengers in this tropical region. Several adult baboons that died from other causes during 1968–69 were largely decomposed and partly eaten by bush pigs and other scavengers in less than 3 days; a dead adult bushbuck discovered in Mkenke Valley was reduced to the skull and a few limb bones in about a week, and even these were gone in 2 weeks. Together with the efficiency of carcass consumption by the chimpanzee, these points are perhaps worth reflecting upon in connection with the scarcity of preserved fossil primates—especially apes—in regions of Africa that are or were environmentally similar to what is now the Gombe National Park.

One of the most conspicuous parts of meat-eating by chimpanzees is the consuming of the brains. In contrast with the high degree of permissiveness observed during distribution, sharing of the brain tissue is guarded and limited. The removal and eating of the brain may be watched closely by others (see Plate 9), but advances and requests for brain are nearly always ignored or discouraged by the individual who has the skull. The brain is the only organ for which marked preference is regularly shown, and the eating of brain tissue is always a slow, meticulous procedure with a definite undertone of enjoyment.

The 7 cases of brain consumption observed in 1968–69—some from less than 2 yards' distance—included 5 baboons, a colobus monkey, and a bushbuck.[19] Mike ate the brains of 4 baboons and the colobus, Leakey of one baboon, and Hugo of the bushbuck (Table VII, Appendix B). Despite the number of chimpanzees and the variety of prey involved, all but one of these cases are quite consistent in terms of technique. In the exceptional case of April 8, 1968, Mike enlarged the foramen magnum to gain access to the brain tissue of a baboon. The remaining skulls were entered through the top of the cranium, and each opening was made at some point on the crest between the frontal bones or at the juncture of the frontals and parietals. A preference for this manner of entering skulls may be inferred from the fact that the technique was used twice when the foramen magnum appeared to be clear of vertebrae and muscles.

In some instances entry is clean and direct, with no accompanying destruction of the surrounding area. But the complete sequence (see Plates 7–11) sometimes begins with puncturing of the skin near the supraorbital ridges, after which the scalp is peeled back over the

19. April 2, 8, and 29, August 7, October 4, and December 23, 1968; and March 11, 1969.

Mike (hand visible) holds the partially consumed skull of a colobus monkey, probably an adult female.

cranium. The face may be skinned as well, and the ears, eyes, and tongue eaten (Plate 22). Pressure is exerted upon the surface of the cranium with the canines and incisors until the bones crack, at times with a force that causes the entire body of the chimpanzee to tremble. Then the aperture is gradually enlarged with the jaws and fingers in an almost perfectly circular fashion to a diameter of about 2 inches—a point at which several fingers can be inserted. The brain is scooped out of its case with the first 2 or 3 digits, and this is usually accompanied by audible sucking and licking. As the brain tissue is removed, the aperture may be further enlarged until the entire top of the braincase has been demolished. If other parts of the carcass are still uneaten at this point, the rest of the head is usually ignored until later.

Thus two species of nonhuman primates living in different environments—the woodland chimpanzee and the savanna baboon—are known to kill prey and to remove brain tissue with their fingers after breaking into skulls through the top of the braincase. These observations are particularly interesting in the light of Dart's (1949, 1953, 1955, 1957) hypothesis, based on study of numerous fractured baboon skulls recovered from australopithecine fossil sites in South Africa, about the bone-tooth-horn implements and inferred hunting behavior of these early hominids. The predatory behavior of chimpanzees living in the Gombe National Park provides a new vantage point from which to conceptualize some aspects of the evolution of primates. So the fact that these chimpanzees capture and kill prey efficiently without using weapons indicates that weapons are not necessarily a definitive feature in the development of hunting behavior among early primates.

Another activity pattern of meat consumption is the "wadging" of meat, and particularly brain tissue, with fresh green leaves from vines, bushes, or trees (fig. 9). This feeding activity, which is also present in conjunction with mastication of other specially favored foods such as eggs and bananas, consists of combining vegetable matter with meat into a pulpy mass, or wadge, which can then be chewed and sucked at leisure. Since the wadge is rarely swallowed, this manner of eating seems to facilitate consumption of unusually soft or liquid foods, or it may be a means of prolonging the pleasure of masticating favored foods. Used wadges may be given to another chimpanzee, or they may be discarded after being thoroughly shredded. However, on April 29, 1968, Leakey was seen to use a leaf wadge to clean out the braincase of a baboon: he inserted the shredded greenery through the cranial aperture, rubbed it around the interior of the skull with his fingers and then took it out and chewed it. A similar procedure, reported previously by Lawick-Goodall (1965), is used to get drinking water from bowls in tree trunks. This is the only instance on record at Gombe of an object being used purposefully during predation.

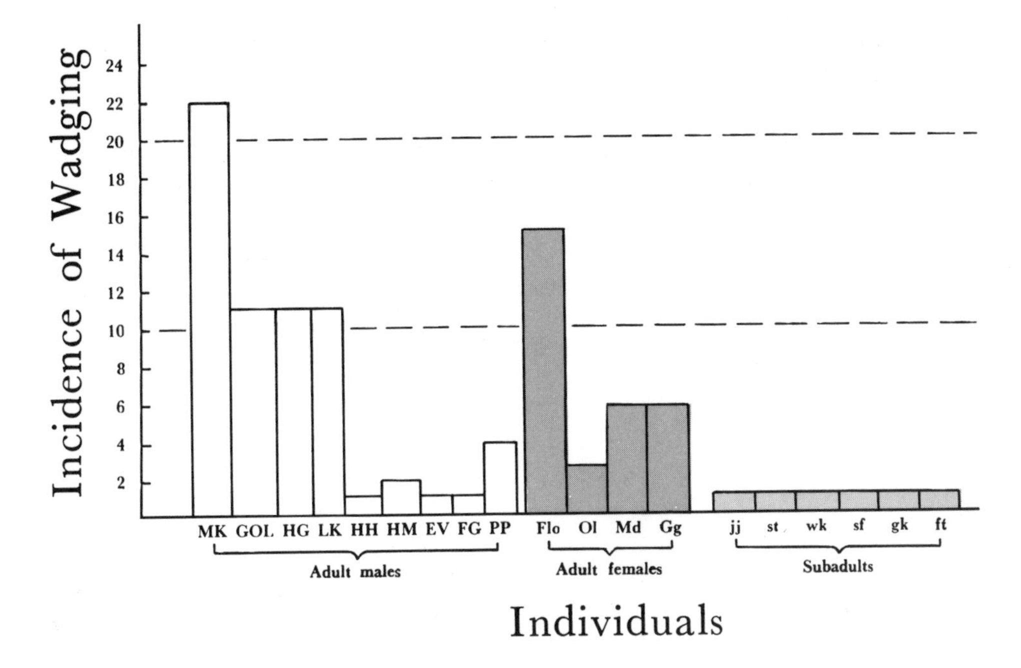

Figure 9. The frequency of meat-leaf wadging (from a total of 100 cases) among individuals of 3 major classes, as observed in 12 episodes during 1968-69. These data may be a rough indication of the relative amount of meat consumed by individuals in different age-sex classes.

b. *Distribution of Meat*

The distribution of meat is by far the most complex aspect of predation. A major part of the field observations are on these activities. Nevertheless, the patterns and social regulators of distribution elude understanding because of highly variable chimpanzee participation in predatory episodes and because of the difficulty of assessing the influence of personality, relationships, and other complicated social factors upon the many and varied interactions observed.[20] Carcasses are usually shared by several chimpanzees through an intricate network of interactions that could not always be recorded fully when several sharing clusters were operating simultaneously. Also, the direction and extent of meat distribution follow complex patterns that differ from episode to episode.

As in all chimpanzee social activities, communication is an integral part of meat distribution. A variety of vocal and gestural signals occurs in connection with the sharing of a carcass. The sounds, gestures, and facial expressions seen during this stage are not specific to predation, though some may appear more frequently in this than in any other behavioral context.[21] Request signals, for example, occur

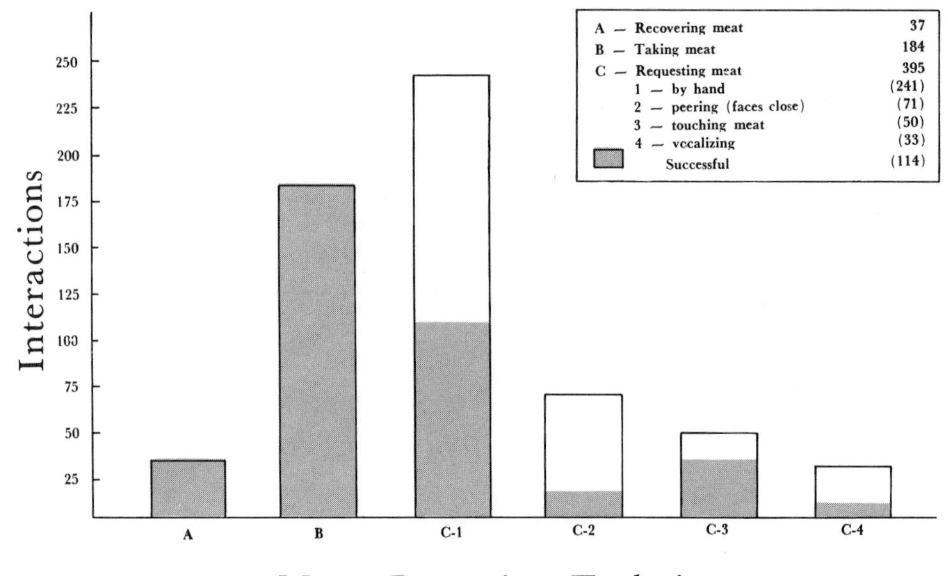

Figure 10. The relative incidence of various modes of obtaining meat during consumption (Stage 3), tabulated from 616 interactions observed in 12 episodes during 1968-69.

very frequently during distribution: 395 begging interactions, containing three to four times as many discrete signals, were observed in the 43 hours of meat consumption during 1968–69.

20. Refer to Appendix C for diagrams of distribution.

21. Following the terminology of Lawick-Goodall (1968b), the common sounds which occurred during distribution were barks, shrieks, and grunts (while feeding); soft barks and waa barks (with aggression); bobbing pants, pant shrieks, and squeeks (with submission); hoos, whimpers, glottal screams, and tantrum screams (with anxiety or frustration, or during begging).

In broad terms, distribution occurs in the following ways: (a) the recovery of meat, (b) the taking of meat, and (c) the requesting of meat. These styles of procurement were observed a total of 616 times during the 1968–69 episodes, and 335 (54%) resulted in the successful transfer of meat from one chimpanzee to another (fig. 10).

Those who do not get meat by interacting are likely to *recover* pieces from the ground. Only small fragments could usually be obtained in this manner because larger portions, such as limb segments, are seldom discarded or dropped. Juveniles, adolescents, and certain adult females are the most likely to concentrate on retrieval of meat, although adult males may do this sometimes.

Meat is *taken* when one chimpanzee removes pieces from the portion, hands or even mouth of another, or when two or more eat from a carcass simultaneously. Such interactions usually occur between adults—mostly between adult males—but infants are sometimes permitted to take meat from mothers, older siblings, and even adult males.

Meat-*requesting,* or begging, is the most complicated form of obtaining pieces because different sign-signal patterns are used in sequential combinations. Paraphrasing Lawick-Goodall (1968b:369), the

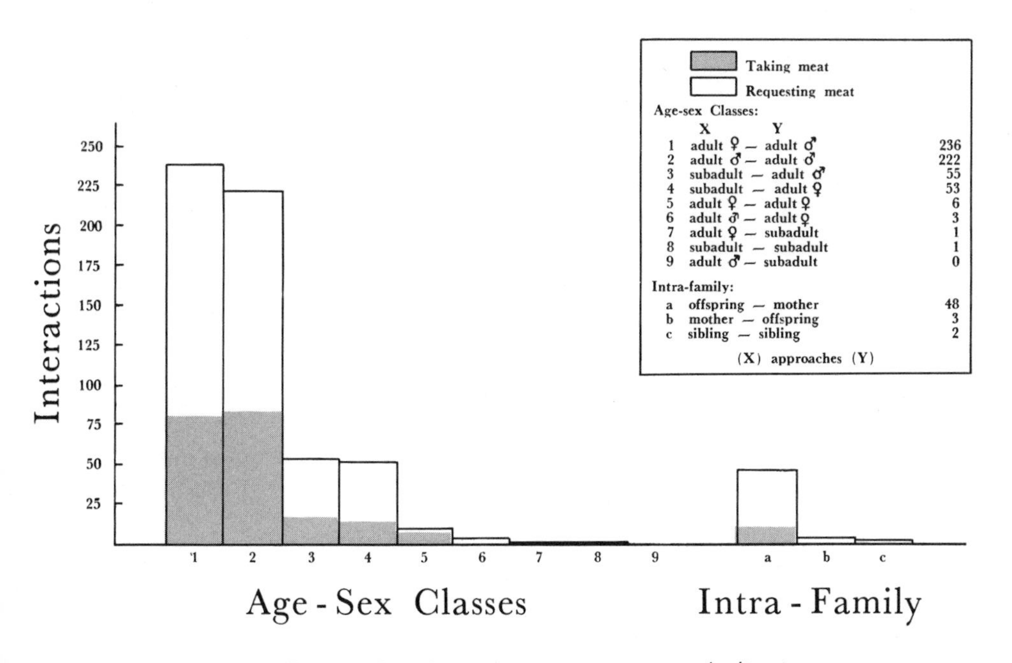

Figure 11. Individual participation in requesting and taking interactions during consumption (Stage 3), tabulated from 577 interactions observed in 12 episodes during 1968-69.

requesting individual may (a) peer intently while placing his or her face very close to the face or portion of the meat-eater, (b) reach out to touch the meat itself or the chin and lips of the meat-eater, (c) extend an open hand, with the palm upward, beneath the chin of the meat-eater, or (d) emit soft *whimper* or *hoo* sounds while doing any of the above (Plate 23).[22] Members of all social and sex classes above the age of about 2 years can participate in requesting meat. Responses to requests may be ignored by simply turning the back toward the begging individual, by pulling away the meat and holding it out of reach, by moving to another location, or by signaling a negation. Positive responses to requests include permitting the begging individual to eat directly from the carcass, allowing him to remove small pieces, dropping chewed pieces into the other's outstretched hand, and actually tearing off a section of meat and handing it to the waiting individual. Positive responses are commonly of the first three types, but the *voluntary handing over of meat* was observed 4 times during the year.[23]

The frequency with which taking and requesting interactions occur varies within age-sex classes (fig. 11). Interactions among adult chimpanzees clearly predominate, as interaction combinations of types 1, 2, 5, and 6—where males or females approach males or females—account for 80% of the 577 available instances. An additional 19% concerns interaction combinations of types 3, 4, 7, and 9—where subadults approach adult males or adult females, and vice versa. But nearly all of these are interactions between infants and their mothers or between infants and adult males with whom the mothers are similarly occupied at the time. Finally, procurement interactions of type 8—where subadults approach subadults—constitute less than 1% of the total. Other interesting combinations of interactions indicate that adult males are much more likely to take and request meat from other adult males (38%) than from adult females (< 1%), whereas adult females take and request mainly from adult males (40%) rather than from other adult females (1%). These comparisons reflect the fact that adult males obtain large portions more often than do members of any other class, and they also emphasize the key role of males in cluster formation and meat distribution.

Considering meat procurement in terms of success in obtaining shares, it is clear that requesting interactions (395) exceed taking interactions (184). But a considerably lower incidence of success in

22. Also see Plates 10, 12, and 17.
23. Refer to Diagrams V, VI, VII, and IX in Appendix C.

PLATE 23

Mike uses the common extended, open-hand gesture in requesting meat from Leakey (foreground). (See also Lawick-Goodall, 1971.) By courtesy of Houghton Mifflin Co.

Mike watches as Leakey begins to move away after refusing to share meat.

requesting interactions (144) largely equalizes these methods of procurement.

As indicated by these data, meat distribution is an involved, active social occasion at which many of those who happen to be present usually participate, and in which nearly all members of the study community took part at least a few times during the 1968–69 period. Participation in sharing varies greatly from episode to episode, due mostly to the variable attendance that results from fluid associations among individuals—that is, in the constantly changing composition of travel, feeding and social groups. With high attendance at consumption sites, meat tends to be more widely distributed and proportionately lesser amounts are obtained by each participant. Also, well attended episodes usually include more chimpanzees who obtain no meat. With limited attendance, on the other hand, distribution is likely to include all participants on a more equal-share basis.[24]

c. *Competition over Meat*

An overwhelmingly large proportion of distribution activity revolves around the individuals who possess large meat portions that are consumed over long periods of time. Overcrowding at these places is somehow regulated, however, because clusters of more than 3 or 4 chimpanzees are seldom seen. But agonistic behavior in and around the meat-eating clusters is rare. Only 82 interactions that could be interpreted as "aggressive" were recorded during 43 hours of consumption, and 68 of these interactions (83%) were restricted to rather mild noncontact interactions in which specific vocal (e.g. *waaing*) and gestural (e.g. arm-waving) threats were most common. The remaining 14 instances (17%) can be classed in the more violent category of attacks because one chimpanzee shoved, dragged, or hit another. Most of the milder cases concerned "objections" to advances for meat, whereas the more vigorous cases appeared to involve the "redirected frustration" of individuals who did not have meat.

Chimpanzees possessing meat portions might register objection by raising the hair, tensing the body, waving the arms, or by vocalizing at others. Sometimes objection is expressed more emphatically by pushing away or lightly slapping the hands or face of the individual attempting to take or request meat. Perhaps because meat-requesting is more frequent than meat-taking, most objections occur in connec-

24. Compare attendance figures with Stage 3 participation figures in Table VI, Appendix B.

tion with the former: only 4 taking interactions (2% of 184) produced a threat response, whereas about 50 requests (13% of 395) were emphatically denied. That most taking is done by a few high-ranking individuals—that is, usually by mature males taking from mature males—may also be reflected in these figures.

Most objections are registered by adult males toward females and subadults. For example, juvenile Sniff was intimidated on many occasions by glares, *soft barks,* and arm-waving from adults because he was especially persistent in approaching clusters for scraps. And infant Flint, whose mother, Flo, was regularly an active participant in clusters, would often be fended off by sparring hand motions or light shoves from adults. Interestingly, much more emphatic objection was sometimes applied by Flint's mother in the form of kicking and beating, especially when continued lack of success in obtaining meat led to *screaming* and thrashing tantrums by the infant.

Persistent refusals are probably the main stimulants of more violent displays and attacks, or "redirected" aggression. These incidents are usually initiated by individuals without meat—mostly adult males—against other individuals who do not have meat—mostly females and subadults. Conversely, combative interactions occur very seldom in direct conflict over meat. An impressive example of the length to which a situation can go without producing an attack was the August 7 episode in which Figan charged Goliath, but without molesting or striking him, and "stole" a portion of the carcass. Goliath objected by *waaing* and stamping on branches and eventually by displaying through the tree, but he did not chase or attack Figan to retrieve the meat. Similarly, on December 3 there was no conflict over the baboon carcass despite Hugo's persistent refusal to reward Mike's requests.

In summary, these observations indicate that agonistic behavior occurs rarely in proportion to other forms of interaction during the last stage of predation. And when aggression does occur it usually appears in contexts other than direct competition over meat.

PREY RESPONSE

Evasion and retaliation responses to pursuit have been discussed in connection with unsuccessful predation, and these are similar in successful episodes as well. Two important aspects of prey pursuit were the absence of injuries among adult chimpanzees and baboons and the comparative inefficiency of protective retaliation by adult baboons. These generalizations also apply during capture and consumption.

Although chimpanzees tend to be silent during pursuit, pandemonium is a regular feature of the capture. Noise and activity were even more intense when many baboons converged upon the scene of capture in all but 3 episodes during 1968–69, and these mixed vocalizations could be heard throughout the valley.[25] Observers learn to identify these sounds with predatory behavior not because the components are different in the predatory context, but because the variety and intensity as a whole is distinct. Chimpanzees can apparently do the same, for these sounds often signal them to converge from many points in the valley. Baboon troop convergence is probably stimulated by specific baboon sounds coming either from the captive or from the adult baboons from whom the prey was taken, for baboons tend to ignore the capture of other prey species even when a troop is in the vicinity.

When many baboons arrive on the scene of capture, a sudden increase of aggressive interaction between the prey and predator species is likely to occur, together with milling movements by individual baboons and chimpanzees. The simultaneous convergence of many excited chimpanzees on the site probably augments predatory success to some extent because a segment of the baboon troop would thus be diverted from harassing and attacking the captor (s). Interactions of the type seen on March 19, 1968, when a baboon leaped onto Mike's back after capture of an infant, were rarely observed.

Baboon response to capture changes rapidly when the captive dies. If the infant baboon is already dead when the troop converges, retaliatory activity is brief and relatively mild. Most baboons calm down and lose interest within 5 or 10 minutes after the kill and depart from the scene within half an hour. But if the captive stays alive, and particularly when it continues to struggle and vocalize while being held, adult baboons may become highly agitated and repeatedly threaten and charge any chimpanzee in the vicinity of the infant baboon. A striking example of prolonged retaliation was observed on April 15, 1968, when adult baboons—mostly males—incessantly harassed numerous chimpanzees around the captured infant baboon, Beau, during the 40 minutes he remained alive. The troop departed within 10 minutes after Beau's death. The effectiveness of harassment was probably reduced on this particular occasion because the chimpanzees took

25. August 7 (colobus kill) and October 4, 1968 (troop absent), and March 11, 1969 (bushbuck kill).

the captive into high trees where access was limited. Withdrawal into trees is, in fact, a useful protective measure against the potentially dangerous adults of various prey species, including baboons, bush pigs, and bushbucks. Even baboons have difficulty approaching the captor (s) when the main access branches are blocked or guarded by individual chimpanzees.

The mothers of captured baboons are sometimes exceptions to the general rule of rapid loss of interest after death of the prey. On several occasions female baboons stayed in the vicinity of the meat-eaters from 2 to 3 hours after their troops departed. These females are likely to watch the activity from a safe distance without interacting with chimpanzees. One mother returned alone to the kill site several hours after her infant had been consumed and wandered all about the area.

Sometimes a different troop of baboons passes near the site of consumption hours after the first troop has departed. These baboons appear to ignore the meat-eaters even when the prey is still recognizable to human observers as being a baboon. In like manner, chimpanzees, even those who have obtained no meat at all usually ignore the new troop. Another attempt at predation has not been observed in these situations.

An interesting and unusual variation on baboon response was recorded on August 5, 1970. Observers arrived on the scene too late to see the capture, but the initial division and the consumption of the carcass were observed in detail. Camp Troop baboons may have captured and started eating the young bushbuck doe before chimpanzees arrived and took the carcass. One adult male baboon stayed particularly close to the carcass even when a cluster of 5 adult male chimpanzees gathered to divide it. This baboon actually joined the cluster for about 10 minutes, sitting side by side with the excited chimpanzees. The chimpanzee members of the cluster tolerated the adult baboon, neither threatening nor chasing him away; the baboon did not openly take meat from the carcass. Several other baboons, including subadults and females carrying infants, also remained nearby, wandering back and forth under the tree where the bushbuck was eventually taken, and foraged for fallen scraps along with chimpanzees who were similarly occupied. Members of both species were in close proximity during the first 50 minutes of consumption, but no aggressive interaction was observed. When chimpanzees later took the carcass to another stand of trees, the baboons remained behind at the kill site for another hour, many searching for fragments on the ground. Several baboons and chimpanzees recovered small pieces. Similar situations—where baboons maintained such prolonged interest in meat, where some baboons in fact obtained meat fragments, and where baboons were so highly tolerated by chimpanzees eating meat—were probably not observed in 1968–69 because the prey were usually young baboons.

An important aspect of chimpanzee predation upon baboons is the absence of severe, injurious retaliation by the prey community, particularly in situations where the predators are outnumbered. No acceptable explanations have been formulated even by those researchers present at many episodes. There is the possibility that the normal retaliatory responses of savanna baboons are somehow inhibited among Gombe baboons when chimpanzees are the predators. Perhaps frequent nonpredatory interaction between members of these species has produced a "pseudo-communal" atmosphere in which baboons are more likely to respond to chimpanzees as they would to other baboons rather than as they normally respond to dangerous extracommunity predators. Accordingly, tolerant and social behavior may dominate and override the normal retaliatory responses when predation occurs. Although these hypotheses relate to many of the anomalies apparent in the 1968–69 episodes, they are too complicated and contain too many assumptions to allow for proof.

SOME SOCIAL REGULATORS OF PREDATORY BEHAVIOR

The complex framework within which any specific set of chimpanzee activities must operate is only gradually becoming clear in the Gombe study community. Discussion of the social context and the controls which regulate predatory behavior is severely limited at present. First, because many specific social factors are unknown, it is difficult to recognize those which regulate predatory behavior. Second, numerical analysis of the various controls which presumably operate during the appropriation and dispensation of meat becomes a formidable scientific enterprise and one that requires a larger sample of interactions than is presently available, because 1225 possibilities for temporary or persistent dyadic relationships exist in a study community of 50 chimpanzees.[26] Consequently, discussion and speculation must focus upon only a few social regulators: those which are mainly the interactions

26. The formula $N(\frac{N-1}{2})$ determines the maximum number of dyadic relationships possible within a community (Carpenter, 1964: 343).

controlled by status, family, and consort relationships among individuals. A perspective of the role of these relationships indicates that predatory behavior is only one of many components comprising a very complex behavioral system. Nevertheless, predation may bring into focus and mirror the wider social structure.

a. Social Status and Control Role

Cooperation during pursuit, mutual tolerance during carcass division, and permissiveness during meat distribution are attributes that deserve special attention because status assertion (or dominance behavior) has been traditionally emphasized by ethologists in terms of aggressive competition over preferred foods. Because meat is a special and highly preferred food among chimpanzees, and because status is a functional derivative expressed in many social contexts, fairly consistent access priority might be expected during predation. This does in fact occur. But priority of access to meat does not conform rigidly to a predictable social pattern, nor is access determined by overt competition.

Inasmuch as adult male chimpanzees are the principal participants in all stages of predation (Appendix B, Table VI), this age-sex class provides the most firm basis for speculation. Adult males can in general be categorized according to top, high, low, and transitional rank (Appendix B, Table II). These categories of status may be in approximate agreement with relative ages, but there appears to be such variation within and between categories that no stable linear hierachy is apparent. The only well-defined, consistent position in the study community appears to be that of the alpha male, who is usually the center of attention in group activities. Although the outcome of agonistic interactions among individuals at the extremes of the rank continuum, such as Goliath and Pepe, may be consistent over a period of years, the results of interactions among closely placed individuals, such as Goliath and Leakey, are likely to vary from one situation to another within short periods of time. Systematic investigation of the network of status relationships is now in progress, and the new information should tell us much about priorities in predatory activities.[27] In the meantime, some speculations about the regulating effects of social status and control role (Bernstein, 1970) during predation must suffice.

During predatory episodes, relative social status may be strongly

27. A special study of adult male relationships was initiated by David Bygott in 1971.

affected by a temporary control role. Control role also operates in a few other contexts, as in leadership of travel groups, but its situation-delimited function is more spectacular in the tense and excited atmosphere of the predatory situation. If an episode unfolds in such a way that a high-status chimpanzee also happens to become the control individual, then interactions are regular and predictable. On the other hand, if a low-status chimpanzee becomes the control individual as a result of acquiring the prey or obtaining a large portion, then other chimpanzees, including those of much higher social status, tend to be displaced from being the focus of action in controlling meat distribution. Thus, although the alpha male, Mike, often acquired the carcass and became a combination alpha-control individual during consumption, he was also observed on several occasions to request meat from a low-status control individual. Moreover, such requests are not always successful, and at times even females or subadults manage to obtain meat from the control individual when the alpha male does not. Observing the entire study community as it was in 1968–69, the lower limit for control role affecting social status during predation was at the level of the young adult or handicapped males, like Figan and Faben, for these were sometimes brazen and at other times prone to intimidation. No female or subadult became the control individual during the 1968–69 episodes.

Examination of the numerical data on participation in carcass division (Appendix B, Table VII) shows correlations between control role and estimated social status (Appendix B, Table II) among the adult males. Correcting the figures to allow for variation in attendance of predatory episodes, individual males can be tabulated in terms of success in obtaining large portions of meat. The control role "sequence" determined this way reflects the categories of social status estimated from other interactive criteria.[28] Four categories of control role among males are represented in the predation data: (a) the alpha and ex-alpha adults, (b) the past prime and prime adults, (c) the physically or socially abnormal adults,[29] and (d) the young adults. The linear arrangement of individuals (Appendix B, Table VII) is not necessarily representative of a linear status hierarchy, but the general scheme does illustrate how predatory behavior may reflect relative social status despite the rarity of overt competition.

---

28. Compare the lists of individuals shown in Tables II and VII, Appendix B.

29. WZ had pathologic and genetic abnormalities which may have interfered with normal physical and social development; RX was a recent immigrant and thus had uncertain rank; and FB and PP were each handicapped by polio paralysis of one arm.

Many of the activities of the alpha chimpanzee, Mike, have been described for the different stages of predation. But the behavior of Mike, who succeeded Goliath to the alpha position in 1964, deserves some added attention in connection with carcass division and meat distribution. Mike was present in 26 of the 30 episodes which occurred in 1968–69, and in 10 of the 12 cases when prey were captured. His participation in all cases was highly active and in most cases central, inasmuch as he frequently initiated predation and acted as a control individual. Yet his high social status did not appear so assertive and functional during predation as it was in other activities because he often permitted others to take meat when he acquired a portion and because his social status was not a guarantee of food reward when someone else was the control individual. Several examples illustrating these points are available.

On April 15, 1968, Mike, together with Hugo, Charlie, and Rix, pursued and captured an infant baboon. However, Hugo acquired the prey, and Mike spent more than 2 hours requesting and taking a few small bits while Hugo consumed the rest. Mike obtained only a small amount of meat and did not threaten or attack Hugo in order to get more. On December 3, 1968, Mike, Hugo, Goliath, Humphrey, Hugh, and Charlie captured another baboon, and Hugo again acquired the carcass. Mike stayed beside Hugo for 2 hours, persistently touching the carcass and requesting meat, but Hugo equally persistently ignored these advances and permitted Mike to obtain only a few tiny fragments. Despite Mike's lack of success—which on this occasion did at times produce a visible degree of frustration—he again did not assert his higher status by forcibly taking the meat from Hugo.

Both of these episodes demonstrate the importance of a control role in regulating meat distribution. Similar behavior is rare in other feeding contexts; bananas, for example, are also limited in supply and high in demand, but competition in accordance with social status usually regulates acquisition and distribution.

Variations on the above theme did occur, however. On April 8, 1968, several adult chimpanzees—Mike, Hugo, Leakey, Goliath and occasionally Worzle, Humphrey, and Flo—together ate from a carcass which was not in the control of any one individual for the first half hour. When the carcass was eventually divided, Mike was not alone in retaining a major portion. In this and similar cases, Mike did not threaten any of the other individuals who participated in the division, nor was there agonistic interaction among the other participants, some of whom differed considerably in social status. On October 4, 1968, Faben managed to retain a captive baboon for several minutes after higher-status males—Mike and Hugo—arrived on the scene. Faben eventually released the baboon after Mike and Hugo placed their hands upon it, but neither harassed Faben or took the prey away forcibly.

Despite the low incidence of aggressive interaction between the alpha male and other members of the study community during predation, estimated social status is nonetheless reflected by carcass division and meat distribution activities. Mike consistently acquired preferred portions like the brains during division and probably consumed a greater volume of meat than any other chimpanzee during the 1968–69 episodes. He consequently rated highest in social status and was most frequently the control individual. This tendency is slightly less apparent but present in the adult female class because Flo, who ranks highest in most social contexts, also took and requested meat more consistently than any other female.

b. Families

Special ties among members of a chimpanzee family—that is, between a mother and her offspring and between siblings—are an important aspect of the social structure. These relationships play a role in assistance during distress, in raising infants, in the adoption of orphans, and in many other contexts.[30] It also is probable that intrafamily relationships are important regulators of meat distribution and may also strongly affect the development and learning of predatory behavior in certain individuals.

Fragments of a carcass rarely pass through more than one pair of hands. The major exception to this rule is the direct passage of meat from mothers to their infants. As mothers of dependent infants are inclined to be more peripheral in many of their daily activities than are other adults, their limited participation in predation, as compared to other mature individuals, is not surprising. Nonetheless, the available data show that the meat procurement activities of mothers and of their offspring are related, with some variation introduced by age differences among infants (fig. 12). Thus, the exceptionally high rate of Flo's participation in meat distribution is mirrored by the participation rates of her offspring, and particularly by that of infant Flint;

30. Intrafamily relationships are now undergoing detailed study by J. van Lawick-Goodall and several assistants.

conversely, Passion's lack of participation is also shared by her daughter Pom.

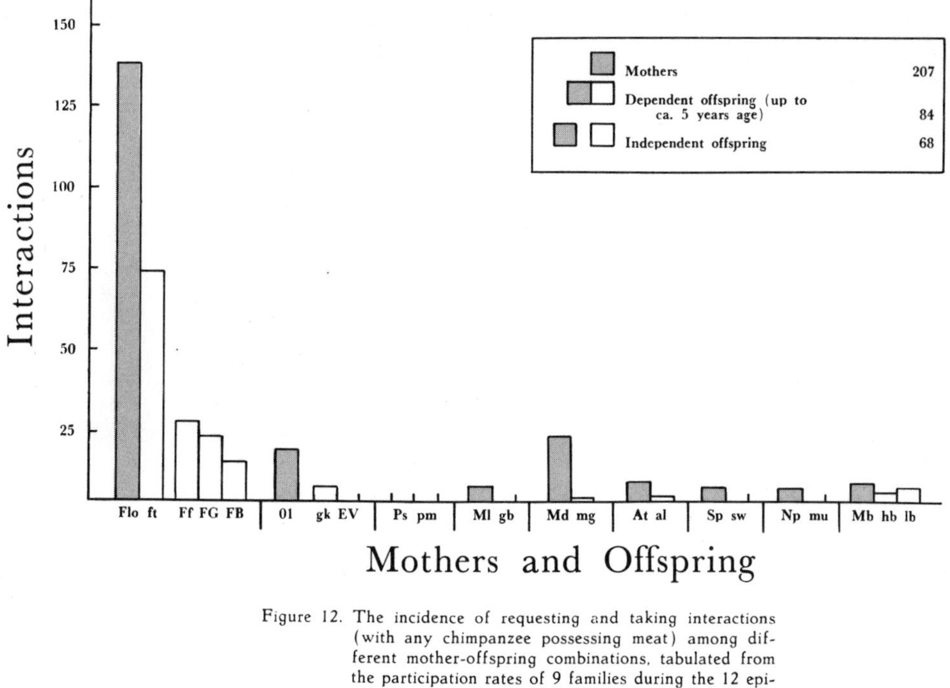

Figure 12. The incidence of requesting and taking interactions (with any chimpanzee possessing meat) among different mother-offspring combinations, tabulated from the participation rates of 9 families during the 12 episodes observed in 1968-69.

The uncommonly large Flo family provides some tentative information about the influence that family members may have upon each other during predation. In the 1968–69 period this family included 6 individuals: mother Flo (age unknown), sons Faben (15–16 years), Figan (12–13 years), and Flint (4–5 years), and daughter Fifi (9–10 years); another daughter, Flame (0–6 months), was born and died during the study period. Among other qualities of this family, relationships among its members appear to be very persistent, and various combinations of individuals often interact and travel together (Plate 24). As shown in Figure 10, participation in requesting and taking meat by the 4 older siblings correlates inversely with their ages. Two factors may be involved in this. First, older offspring are increasingly less likely to obtain meat from their mothers, so Faben, Figan, and Fifi had to rely, much more so than Flint, upon taking or requesting meat from adult males who possessed large portions. Second, the correlation between a rise in age and a decrease in procurement interactions may reflect another correlation between a rise

PLATE 24

Intra-family grooming—Fifi on Flo on Figan—as infant Flint wrestles with another infant (background).

(Above) Mother Flo (far right) and her ad daughter Fifi (far left) groom Flo's newest fant, six-month-old Flame, while son Flint (c ter) watches closely. (Below) When Flo tires grooming Flame, Fifi pulls the infant away order to groom more intensively.

Flo travels with two infants, Flint (4.5 years) riding dorsal and Flame (3.5 months) clinging ventral.

When Figan (far left) showed distress or apprehension by raising his hair and grimacing, his older brother Faben (far right) joined him and did the same; the two adult male brothers were joined in turn by their youngest brother, Flint (center), who embraced Faben as all three faced toward a cluster of adult males (off camera) nearby.

in age and a decrease in the "protective aura" of a mother's social status. In simpler terms, the preferential treatment given by other adults to a mother may assist her offspring less and less in obtaining meat as the age of the young increases. The high incidence of taking and requesting by Flint as compared to the low frequencies exhibited by Fifi, Figan, and Faben are probably indicative of these intrafamily relationships.

Looking at the same data from a slightly different point of view, Figure 10 also shows that participation rates tend to be more uniform within a family than between families. All of Flo's offspring are more active in taking and requesting meat than are any other offspring of similar ages. More specifically, Figan was more active in all stages of predation than was Evered during 1968–69, despite the fact that both were in the same age-sex class. These differences in families may indicate that entire family units, not just individuals, have social status.

Aside from the still puzzling and often ambiguous problems posed by family relationships in regulating meat distribution, there are further questions concerning the part played by learning in any close relationship. Again referring to the Flo family, it is clear that her offspring, all of whom were probably exposed to predatory behavior with some regularity when they were still subadults, now participate more consistently than other chimpanzees of comparable maturity. Flame, born in August of 1969, began her exposure to predatory behavior in December of that same year, at a mere 3 months of age. By the end of 1968 she was watching meat-eating activities at close range from her ventral position on Flo. Similarly, Athena's son Atlas was observed taking and eating meat when less than 2 years old and still suckling. Another infant born after the 1968–69 study period was inquisitively touching the meat in her mother's mouth within her first year. And Flint, by way of his constant association with Flo, was present at predatory episodes during 27 of the 47 net observation hours logged in 1968–69—fully 20 hours more than any other infant in the study community. Combined with the fact that all of Flo's offspring were relatively frequent participants in predation, these observations on the early exposure of infants to predatory behavior suggest that learning may have considerable effect upon the amount of participation shown during adulthood.

Turning to the study community in general, mother-offspring and sibling relationships may be an important factor among other combinations of adult individuals whose family origins remain obscure.

This may explain the marked preference that Charlie, for example, shows for requesting meat, and his greater success in obtaining it, from Hugh, who is believed to be his older brother. The regulating effects of many relationships remain uncertain because they cannot be traced during the 10-year study of a primate that has a life span of 40 or more years in natural conditions.

c. Female Sexual Cycles

A variety of intricate relationships, both temporary and persistent, appears to exist between adult male and adult female chimpanzees, but most of these are not yet clear.[31] The effects of these relationships upon meat exchange between these classes are, as a result, difficult to ascertain. However, it does seem clear that female chimpanzees who are in estrus when predation occurs are likely to participate more frequently and to obtain more meat than when they are not in estrus. Adult males are, in general, more tolerant and permissive with estrous females during meat consumption.

One or more estrous females were present in 8 of the 10 episodes where the interactions of meat distribution could be observed in detail during 1968–69, but 3 or more estrous females were present in only 2 episodes (Appendix C, Diagrams I through X). In addition, of the total number of individual females (29) that interacted with males during the year, more were nonestrus (17) than estrus (12). Although these figures show that at least one estrus female is likely to attend each predatory episode and that nonestrous females outnumber estrous females, the regulatory effect female sexual receptivity has upon meat distribution can only be examined by a comparison of the interactions of estrous and nonestrous females with adult males.

In 236 interactions of adult females with adult males (see fig. 11), there were 132 instances (56% of 236) where estrous females took and requested meat from adult males and 104 instances (44% of 236) where nonestrous females did the same. Combining these interaction figures with the attendance figures given above, it becomes clear that an estrous female interacts more with adult males than does a nonestrous female.[32]

A more meaningful measure of the value of being in estrus during predation is given by the success rates for both estrus and nonestrus

31. A special study of female sexual behavior was initiated late in 1968 by P. R. McGinnis, and the material is now undergoing analysis.

32. 236/29=8 interactions per female; 132/12=11 interactions per estrous female; and 104/17=6 interactions per nonestrous female.

interactions in taking and requesting meat. In the 132 instances of estrous females interacting with adult males, meat was obtained 91 times (69% success); whereas in 104 instances involving nonestrous females, meat was obtained only 42 times (40% success). Thus, sexual receptivity correlates both with participation in episodes and with success in obtaining meat.

In more qualitive terms, estrous females stay closer to adult males during predation and are more persistent in their efforts to obtain meat. One cycling female, Fifi, was in estrus during 5 of the 7 consumption periods she attended in 1968–69. She persistently took and successfully requested meat from adult males during 4 of these episodes. Unaccountably, she took little interest in the prey during the fifth episode. In the 2 episodes Fifi attended without being in estrus, she took a small fragment only once and did not request at all from the males possessing meat. Other females did not attend episodes nearly so frequently while in estrus, but the above pattern was repeated on a smaller scale. Some females attended several episodes during the year but only participated when in estrus. The only time that a female was observed to acquire a fair portion during division was on October 4, 1968, when Gigi, who was in full estrus at the time, ate most of the viscera.

All of the social regulators that have been discussed probably operate both independently and jointly. For example, a high-status estrous female would presumably be more tolerated by adult males than a female who shows neither or only one of these attributes. The relative value of the various factors is still uncertain, and the importance of each factor is probably further affected by which chimpanzees are involved. Flo's participation did not appear to be inhibited when not in estrus, which may mean that high social status or special relationships are most significant in some situations. It is impossible to say whether Flo would or would not be more tolerated than a low-status female who was in estrus.

There are, of course, many other dyadic relationships which might regulate meat sharing within the study community, but a better understanding of regulating factors is dependent upon a more detailed picture of the full behavioral repertoire of the Gombe chimpanzees. This goal has been undertaken by members of the Gombe Stream Research Centre, but the task is immense and many years away from completion.

4

SUMMARY

Although 132 cases of predation have been documented during a decade of field research with a study community of 40 to 50 wild chimpanzees (*Pan troglodytes*), the bulk of the material presented in this report on predatory behavior was gathered during a 1-year period —from March of 1968 to March 1969—at the primate station in Gombe National Park, Tanzania. Occupying an 10–12 by 2–3 mile segment of the narrow geographical zone bounded on the west by the shore of Lake Tanganyika and on the east by the escarpment of the Great Rift Valley, the park comprises an area of 25–30 square miles of rugged terrain. Covered by a patchwork of grassland, woodland, and forest vegetation, the region supports a rich variety of small wildlife and a few larger species such as leopard, bush pig, bushbuck, and buffalo. Isolated from popular tourist routes, limited to access by boats and footpaths, and devoid of permanent human settlement, the park is ideal for the undisturbed study of at least 7 indigenous primate species.

Reflecting its gradual expansion since J. Goodall's first field expedition in 1960, the present Gombe Stream Research Centre (P.O. Box 185, Kigoma, Tanzania) provides quarters and other facilities for numerous research and support personnel. A broad study program is oriented primarily toward the naturalistic behavior of chimpanzees but also focuses upon the ecology of the park and the behavior of baboons and other monkeys. In combination with limited banana feeding, a procedure which has been successfully used since 1962 to speed habituation of chimpanzees to human beings, a strict policy of noninteraction between human and nonhuman primates, has promoted a high degree of mutual tolerance that permits close observation of chimpanzees on a daily basis. Because observation of chimpanzee predatory behavior was only one phase of the general long-term program directed by J. van Lawick-Goodall, several members of the research team participated in documenting these activities during the normal course of their varied duties. Many questions and problems raised by these extensive but generally oriented observations on predation require additional field study utilizing a more systematic approach.

An unprecedented 47 net observation hours on chimpanzee predatory behavior were accumulated during the 1968–69 period of field study. Representing a pool of data collected by some observers closely acquainted with chimpanzees and others with baboons, this journal covers 30 episodes involving 28 baboon prey, 1 colobus monkey, and 1 bushbuck. Prey taken in other years include redtail and blue monkey and bush pig. The 30 episodes are a small part of the 132 predation cases recorded since J. Goodall observed the first kill in 1963. Predatory behavior has been observed in only 83 cases of which 46 were kills; analysis of feces and carcass remains accounted for another 49 kills. The 56 prey that could be identified included 6 species of monkey, antelope, and pig, and these comprise nearly all the mammalian species potentially available as prey in the park region. A high proportion of the 1968–69 episodes was observed in the vicinity of the feeding area situated some 700 yards inland along Kakombe Stream, in part because this area received the most constant observational coverage. Although it is highly unlikely that these chimpanzees have acquired predatory tendencies in recent years as a consequence of human intervention, the high incidence of predatory behavior seen during this one year may be related to the recent habituation of two baboon troops to human beings and to the feeding area. Most probably, however, these factors emphasized predation upon baboons in particular during 1968–69 and did not greatly alter the yearly rate of chimpanzee predation.

Three primate units—the chimpanzee study community (membership 48) and two baboon communities (membership 75 and 55)—were the principal participants in the 30 episodes observed during 1968–69. Chimpanzees pursued members of these troops on 28 occasions: Camp Troop was involved 16 times, Beach Troop 6 times, and 6 cases remained uncertain. The predatory success rates upon the 2 troops were 7% and 9%, respectively. In Beach Troop this figure constituted 29% of the subadults below the age of 2 years, which is the approximate upper age limit for baboon prey. These troops ranged over partially overlapping areas containing separate core zones and sleeping sites, while chimpanzees in the study community used a much larger range that incorporated those of both baboon troops. Unfortunately, no comparable data are available on rates of predation upon other baboon troops within the range of these same chimpanzees, or upon troops in other areas frequented by unfamiliar chimpanzees. Predation is known to occur throughout the park, but demographic data on the various prey species do not yet exist.

Observations on interspecific relations among the various nonhuman primates inhabiting the park point to mainly tolerant coexistence. Several combinations of species have been seen in close proximity without showing alarm or antagonism, and some species regularly associate with others. Chimpanzees have been observed while temporarily sharing locations, such as feeding sites, with most of the other primate species. A variety of interspecific social interaction supplements the general atmosphere of tolerance. Chimpanzees on occasion interact with other primate species, such as the redtail, and young chimpanzees frequently play with young baboons. In fact, relations between the 3 study communities—the 40–50 known chimpanzees and the 120–140 known baboons—continued to be fundamentally tolerant even within the context of repeated predation during 1968–69. The anomaly presented by the occurrence of predation that does not significantly interfere with tolerant interspecific coexistence is perplexing and far from solution. These phenomena also raise several questions concerning reports about chimpanzees living in other regions of Africa where similar observations on interspecific tolerance have been considered an indicator of the absence of predatory behavior.

Nonhuman primates throughout the world have long been noted for their role as prey to other species, especially man. Conversely, little reliable evidence accumulated before the 1960s concerning nonhuman primates as predators, and the widespread assumption has consequently been that predatory behavior does not occur, or at most occurs only very rarely in especially "opportune" situations. Evidence to the contrary is gradually emerging from field studies on various continents, and the probability is rapidly increasing that the predatory proclivities of primates in general have been considerably underestimated. The Gombe data are particularly significant because they indicate a previously unsuspected regularity and frequency in predation within a single chimpanzee population. Other field studies which incorporate habituation of *both* the predator and prey species may in the future yield similarly interesting results.

The conditions which precede predatory episodes are not clearly defined, the greatest problem being that analysis is limited to chance details abstracted from general notes. However, the potential effects of such factors as variations in climate and food resources, the disparate availability of prey in terms of geographic distribution and communal concentration, and periodic changes in banana feeding procedures have been outlined. The only consistent factors seem to be the size of prey individuals, which are within 5–20 pounds for all species, and certain vocalizations of subadult baboons. More probing speculation, such as the possibility of adult chimpanzees being more reluctant to prey upon adult baboons, with whom they interacted sociably as subadults, than upon the less familiar younger members of a troop, are intriguing but as yet factually unfounded.

The 30 episodes recorded in 1968–69 are not a sufficiently firm information base from which to formulate final conclusions, but some patterns of predatory behavior are clear. Strictly as a food-getting activity, predation may be viewed as a series of behavioral stages: *pursuit, capture,* and *consumption.* Each stage is characterized by separate and distinct activities, and each can be further segmented into discrete components.

Pursuit can occur in several basic modes: the *seizing, chasing,* or *stalking* of prey. These yielded a ratio of 7:11:4 respectively, if the 8 indefinite episodes during 1968–69 are excluded. Mode–1 refers to "opportune" situations where prey are near enough to be seized suddenly with a minimum of effort. Mode–2 and mode–3 refer to cases requiring greater effort and control, and at times active cooperation, by the predators. Pursuit is apparently initiated and conducted primarily by adult male chimpanzees, as females and subadults have rarely been observed to participate until consumption begins. Indications are that status, which plays a significant part in social contexts, may be temporarily suspended during pursuit, or perhaps equalized by a more immediately important common intent. Thus, social status does not necessarily correlate positively with predatory initiative or leadership. Temporary control role may be a useful explanatory concept in the study of predatory behavior.

Predatory episodes can be classified as kills or attempts. A ratio of 2:3, or 40% successes and 60% failures, was recorded during the 1968–69 period of study. Only 37 attempts upon prey, most of them baboons, have been recorded in 10 years, probably because it is easy for observers to miss silent pursuit if there are no results. Several sets of factors appear to be involved in failure: loss of "interest" by the predator, interference of adverse environmental conditions, evasion by the prey, and retaliation by the prey individual or community. These factors may operate singly or in combination, and they may be supplemented by other factors that arise in specific situations.

When pursuit is successful, the ensuing capture is often dramatically characterized by explosive bursts of predator and prey vocalizations and interactions. This stage consists of several sequential components: the *acquiring, killing,* and initial *dividing* of prey. Captives are usually dispatched by repeated smashing against a hard surface, by throttling or a rapid bite on the neck, or by being literally torn apart if several chimpanzees acquire the prey simultaneously. The use of these techniques can be situationally determined, but consistency among some adult males suggests that experience and learning are also involved. Initial division of the carcass usually occurs at the time of capture, with 2 or more chimpanzees obtaining large portions. Division is included in the second rather than the third stage of predation because the actions are more a matter of appropriation than of dispensation. Appropriation does not imply combat, however, even during the tension and excitement surrounding division of a small carcass by as many as 5 adult male chimpanzees. Competition over meat occurs very rarely. Rights to meat are determined mainly by establishing possession, or ownership. Thus, the sooner an individual obtains meat, and the longer he maintains possession by moving away or hoarding it, the less likely are other chimpanzees to attempt appropriation.

After carcass division comes the stage of consumption, during which meat is shared among many members of the group. Three major components of consumption have been discussed: the techniques used to dismantle carcasses, the procedure whereby meat is distributed, and the nature of interactions focusing on meat. By far the longest stage of predation, consumption begins with the formation of small clusters, most often containing 3 or 4 individuals of diverse ages and both sexes, around those chimpanzees who manage to obtain major portions. These clusters, several of which often operate simultaneously within a radius of a few dozen yards, are the focal points for the complex network of interactions that revolve around distribution of meat shares among as many as 15 individuals per carcass.

Meat distribution occurs in ways that reflect an increasing intensity of effort: the *recovery* of meat from the ground, the *taking* of meat from someone's portion, and the *requesting* of meat from another individual. These means of obtaining shares were observed 616 times during 43 net hours of consumption, and 335 (54%) were successful. Requests include several distinct communicatory signals, none of which is unique to the predatory context. Three major age-sex classes —adult males, adult females, and subadults—participated in meat procurement interactions to varying degrees: fully 80% of 577 interactions occurred among adults, 19% between adults and subadults, and only 1% among subadults.

Displays, chases, and occasional attacks among chimpanzee participants do occur during consumption, but most of these involve individuals without meat who are expressing frustration. Only 82 aggressive interactions were recorded during the 43 hours of observed consumption in 1968–69, and 68 of these (83%) consisted of noncontact threats that were usually objections to advances for meat portions. The remaining 14 instances (17%) were more severely aggressive, but none was direct combat over meat. On rare occasions meat was actually handed by one individual to another in a voluntary manner.

Neither implements nor projectiles have been used by Gombe chimpanzees in connection with the pursuit or capture of prey. A single instance of object usage was recorded during consumption in April 1968, when an adult male chimpanzee cleaned out the braincase of a baboon with a wadge, or leaf sponge. The technique of eating brain tissue by opening a small hole in the cranium, then scooping out the brains with the fingers, is also of interest because it occurs fairly consistently and because it may pertain to damaged baboon skulls recovered from African fossil sites.

The lines along which meat is distributed within the group, or that segment which happens to be in attendance, appear to be regulated by many factors, only some of which have been recognized. Both temporary and persistent relationships among individuals of all ages probably regulate distribution on the bases of social status, family membership, female estrous cycling, and other factors. More subtle and complex (e.g. tripartite) sharing patterns, such as may occur within a family unit or between a male and a preferred female who is not in estrus at the time, are only beginning to emerge from the data. As shown by markedly different rates of participation among members of the subadult class, the extent of participation during adulthood may be affected by degree of dependence upon the mother, introduction to predation at an early age, prolonged exposure through a parent's participation, and probably other variables as well.

DISCUSSION

Despite the volume of information now available on the predatory behavior of chimpanzees living in the Gombe National Park, the Budongo Forest, and the Kasakati Basin area, some aspects of this food-getting activity remain enigmatic. A few of the more obvious characteristics have received detailed attention in previous chapters, but many important problems have been mentioned in cursory fashion. Hence, a few basic queries are selected for informal commentary and speculative discussion, with the hope that this might provide impetus and basis for further observation, inquiry, and contemplation.

a. *Why Do (Gombe) Chimpanzees Practice Predatory Behavior?*

Chimpanzees living in the Gombe National Park, and probably elsewhere as well, consume a wide variety of foods, many of which are available in accordance with fluctuating climatic and reproductive cycles. But the diversity of materials consumed would seem to negate any probability of excessively lean periods during any single year or period of years because many combinations of food types are available at all times. For example, of the different species of edible plants comprising the main food resources, so many are available every month of the year that dependency upon any one, or upon any small combination, is probably negligible. In some seasons, consumption of plant foods may be considerably reduced even by choice because some other preferred food, such as termites, can be readily obtained in large numbers. Logically then (Gombe) chimpanzees probably have sufficient food, in terms of both bulk and variety, for hunger to be discounted as a prime motivator of predatory behavior. That predatory episodes often occur directly after feeding on other foods, notably bananas, supports this assumption.

There remains a possibility, however, that a form of "specific" or "residual" hunger could motivate predatory behavior through differential satiation. Consumption of large volumes of fruit, many of which contain an abundance of carbohydrates, may leave a need or desire for other substances, such as vitamins and minerals, that are more concentrated in fresh meat and bone. This "residual" hunger, which can be expressed behaviorally in a sudden urge to eat fresh meat after steady consumption of vegetable matter, could conceivably stimulate predation.

A problem arises in that certain tentative factors become disproportionately vital to this line of inquiry: (a) that (Gombe) chimpanzees have developed a dietary preference or acquired a physical need for nutritive substances obtainable only or mainly through predation, and (b) that the development of such a habit would require an amount of time that would, in turn, indicate a long history of meat-eating for these apes. Neither of these points can be completely clarified in the field, but controlled testing of dietary preferences and habits in confined chimpanzees might provide pertinent information.

Focusing more closely upon the environment, an explanation of the existence of predatory behavior may lie in the ability of Gombe chimpanzees to fill, perhaps by means of a plastic behavioral capacity, a relatively vacant niche within an ecosystem that has few large carnivores. Observations show that these chimpanzees range throughout their habitat with marked indifference to potential dangers from other predators, moving on the ground through dense forests, thickets, and tall grasses which cut visibility to a few feet. Moreover, the only large carnivores in the Gombe park, leopards, probably play a minor role in reducing the chimpanzee population because no signs of leopard predation have been encountered and because few chimpanzees have disappeared unaccountably during 10 years.

Application of this hypothesis does not ascribe conscious or deterministic motives to an expansion of food-getting activities. One could speculate that (a) chimpanzees which are rarely preyed upon or even intimidated by carnivores would be in an optimum position, even without specific pressure from the environment, to expand an already extensive capacity for utilizing diverse food resources from the level of the primary consumer to the added level of the secondary consumer, and that (b) chimpanzees inhabiting a region containing few carnivores can begin to incorporate this additional food resource, due in part to a presumably high density of prey in such areas, without risking the competition that would prevail in sharing this resource with other predators which might eventually turn to preying upon the chimpanzee population itself.

Causal relationships between physiological or environmental conditions and predatory behavior are in much need of further investigation. At best, general indications are that a diversified and plastic behavioral capacity that operates within a heterogeneous environment may provide the most reasonable base for theories about the origin and development of predatory behavior among primates. Such a contention—which in fact merely suggests that meat-procuring and meat-eating habits may have originated in the varied conditions of woodlands or riverine forests rather than the open savanna—would dispense with assumed prehistoric migrations from forest to grassland environs in order to account for a transition from foraging to hunting behavior by an ancestral primate.

Another plausible line of speculation concerns the possibility that social factors, rather than physiological or environmental ones, serve to reinforce predatory behavior in (Gombe) chimpanzees. The basic elements of this hypothesis are that (a) considerable energy is expended, and often wasted in failure, on acquisition of fresh meat when numerous other foods can be regularly obtained with much less effort; that (b) meat is usually eaten and shared in a leisurely manner more suggestive of pleasure than basic hunger alone; and that (c) small carcasses are usually too widely distributed to provide a filling meal for any one individual. Qualitative observations such as these are no proof of social reinforcement, but they do suggest that solution of the central question may be neither obvious nor simple.

Status is an important ingredient of chimpanzee social life for all age-sex classes. Young males that are in transition from adolescence to adulthood in particular strive to achieve social status. These individuals repeatedly test themselves against more mature and higher-ranking individuals. Now, if control role is in fact functional during predation, then it is plausible to believe that an individual, and especially a low-ranking adolescent or young adult male who is in transition, may in the long run benefit from achieving a temporary control role. A series of such achievements could conceivably raise the social status of the individual because the cumulative effect of repeated control role activities might translate into higher social status. This adjustment may be particularly effective in establishing the relative status of an individual within his own age-sex class, which will eventually replace the older generation entirely. It is also possible that this behavior rewards communal integration and cohesion by equalizing individual differences in age and rank during predatory episodes.

b. *What Factors Contribute to the Predatory Capability of (Gombe) Chimpanzees?*

Chimpanzee predatory behavior is usually a collective activity that incorporates shared objectives and rewards. A solitary chimpanzee has not yet been observed to pursue, capture, and consume a prey. In fact, groups of protagonists are likely to be involved in predatory episodes, inasmuch as baboons and colobus monkeys, both of which are communal species, were the most frequent prey in the Gombe park. Thus, it may seem reasonable to assume that predatory capability is somehow related to collective action. Similar reasoning has been applied to hyenas, wolves, humans, and other group predators that are considered more capable at hunting collectively than individually. This premise may apply to chimpanzees, but it tends to disregard the many differences in styles and conditions among the various predator species. Moreover, head counts in the 1968–69 episodes at Gombe did not consistently show that the total number of chimpanzee participants relative to the total number of baboons in the same area affected the initiation or the success of predation.

Solution of the problem is somewhat blurred by fluctuating participation rates among chimpanzees, and by the fact that some individuals elect not to take part in any stage of predation even when they are present. It is only possible to speculate that relative numbers may be significant in subtle ways: (a) the likelihood of uneasiness, then panic among prey clusters at the start of pursuit, and of eventual confusion among converging members of the prey community during pursuit and capture, may increase in proportion to the number of chimpanzees being confronted; (b) the capability of chimpanzees to keep track of a targeted prey individual probably decreases in relation to the number of prey protectors converging upon the scene of pursuit or capture.

Another important factor in predatory capability may be the multidimensional mobility of chimpanzees. Unlike many large carnivores, chimpanzees possess locomotor versatility in all zones of their environment. Those prey which habitually seek safety from other predators by climbing trees often cannot find sanctuary from chimpanzees, particularly when the latter conduct collective pursuit. Adaptation to multiple habitat zones presumably increases the predator's advantage during pursuit and capture because prey can be approached and pursued in all directions.

This line of conjecture leads to new possibilities for the setting in which predatory behavior may have developed in primate evolution. Protohominids, who presumably possessed more generalized locomotor abilities and adaptations than does modern man, might have found predation far more difficult on the open savanna than in woodland or forest environs where their capabilities could be used to full advantage. An important question then arises about what necessitates the assumption of a habitat change or migration—from forest to grassland—in order to explain the origin and development of predatory behavior among primates. This point is particularly pertinent in view of the fact that some additional "human" characteristics which have in theory been associated with a habitat change (e.g. nuclear family, bipedalism, food-sharing, and tool-making) appear to have their incipient counterparts among living chimpanzees.

c. *Do (Gombe) Chimpanzees Qualify as True Predators?*

This issue remains obscure mainly because no relative or absolute scale or predatory rates has been devised for comparing the meat-procuring and meat-eating habits of mammals in general, nor of primates in particular. Nonetheless, the question is relevant in that a variety of important assumptions, many of which relate to the presence or absence of consistent predatory behavior, constitutes the foundations for several current theories of primate evolution and, more specifically, of human origins.

As a result of the growth of interest in the naturalistic behavior of nonhuman primates during the 1960s, meat-eating has now been observed among members of four taxonomic families of extant primates—namely, the Cebidae, Cercopithecidae, Pongidae, and Hominidae. However, it is still possible in many cases to debate whether meat-eating is a matter of collection or predation. Whatever opinions may be held on this issue, it seems clear that a spectrum of food-getting behavior exists in the Order Primates, with a range of variation extending from the true vegetarian through the occasional meat-collector to the part-time predator. In terms of a general view of primates, certain leaf-eating monkeys would seem to occupy one end of the spectrum, and certain human hunting groups the other end. The (Gombe) chimpanzee probably occupies a point on the spectrum close to the human extreme. Just as human populations occupy a band of variability on this scale, chimpanzee populations are likely to vary similarly, so a final evaluation of the above question must await additional field results from different regions.

Direct comparisons between chimpanzees and other species of predators would be a more useful measure of regularity than the above speculations, but study methods and conditions are often so different that data cannot be meaningfully compared. There is limited value in stressing that (Gombe) chimpanzees are in some circumstances more successful predators than are lions, or in postulating the numerical effects of chimpanzee predation within a given region and then aligning these results with the kill rates of various carnivores. Such calculations and comparisons provide a general impression about the regularity and efficiency of chimpanzees as predators, but there are still too many gaps in the picture. The part-time predatory activities of (Gombe) chimpanzees may not even be in the same behavioral category as the activities of more specialized predators, including such communal species as hyenas and wolves, for the simple reason that these apes are not dependent upon killing prey for their sustenance.

A closer analogy to the predatory behavior of (Gombe) chimpanzees might be expected from ethnographic studies of human hunting groups, but comparison would again be premature because numerical data on the regularity and efficiency of hunting are very limited. Without such comparative data, discussions of the similarities and differences of chimpanzee and human predatory behavior cannot be made. It might, for example, seem reasonable to consider prehistoric hunters, and even some of the hunting communities of today, as more capable and more regular predators than (Gombe) chimpanzees simply because the latter do not kill large mammals. But if human hunters capture few large prey in comparison to the total number killed per year, then some human groups may well have lower success rates than some chimpanzees.

As a final note, it seems appropriate that an agreement was reached during a recent anthropological symposium on "Man the Hunter" to retain the term *hunter* "despite the fact that the majority of peoples considered subsisted primarily on sources *other than meat*—mainly wild plants and fish" (Lee and DeVore, 1968:4). As the regularity of predation by human hunting groups is currently as uncertain as the regularity of predation by chimpanzee groups, there would be limited utility in using such terms as *hunter* and *predator* to distinguish between patterns of food-getting behavior in human and nonhuman primates by implying quantitative differences where only qualitative ones exist.

The range and volume of information collected during 10 years of research at Gombe National Park would seem to reasonably establish these particular chimpanzees as more than occasionally interested or "opportunistic" collectors of meat. The Gombe chimpanzees are neither true vegetarians nor true carnivores, but they may qualify, on the general scale of food-getting behavior exhibited by mammals, as omnivorous forager-predators. As such, these apes may have long ago approached, or are now in the process of approaching, a position in primate evolution that has, in theory, been reserved for prehistoric and modern communities of human hunter-gatherers. Awareness of occasional predatory behavior in several primate species, and of recurrent predation upon a variety of mammalian species by at least one nonhuman primate population, will hopefully stimulate further research and debate about human origins and primate evolution.

APPENDICES

A. MAPS

Maps I and II are to scale, adapted from official maps of the Tanzanian Survey. Maps III, IV, and V are highly approximate, having been drawn from hand sketches that were originally produced by visual examination and pacing of the chimpanzee study area. Aerial photographs of the park region, taken in early 1970 from a low altitude, were used to check features and relative distances.

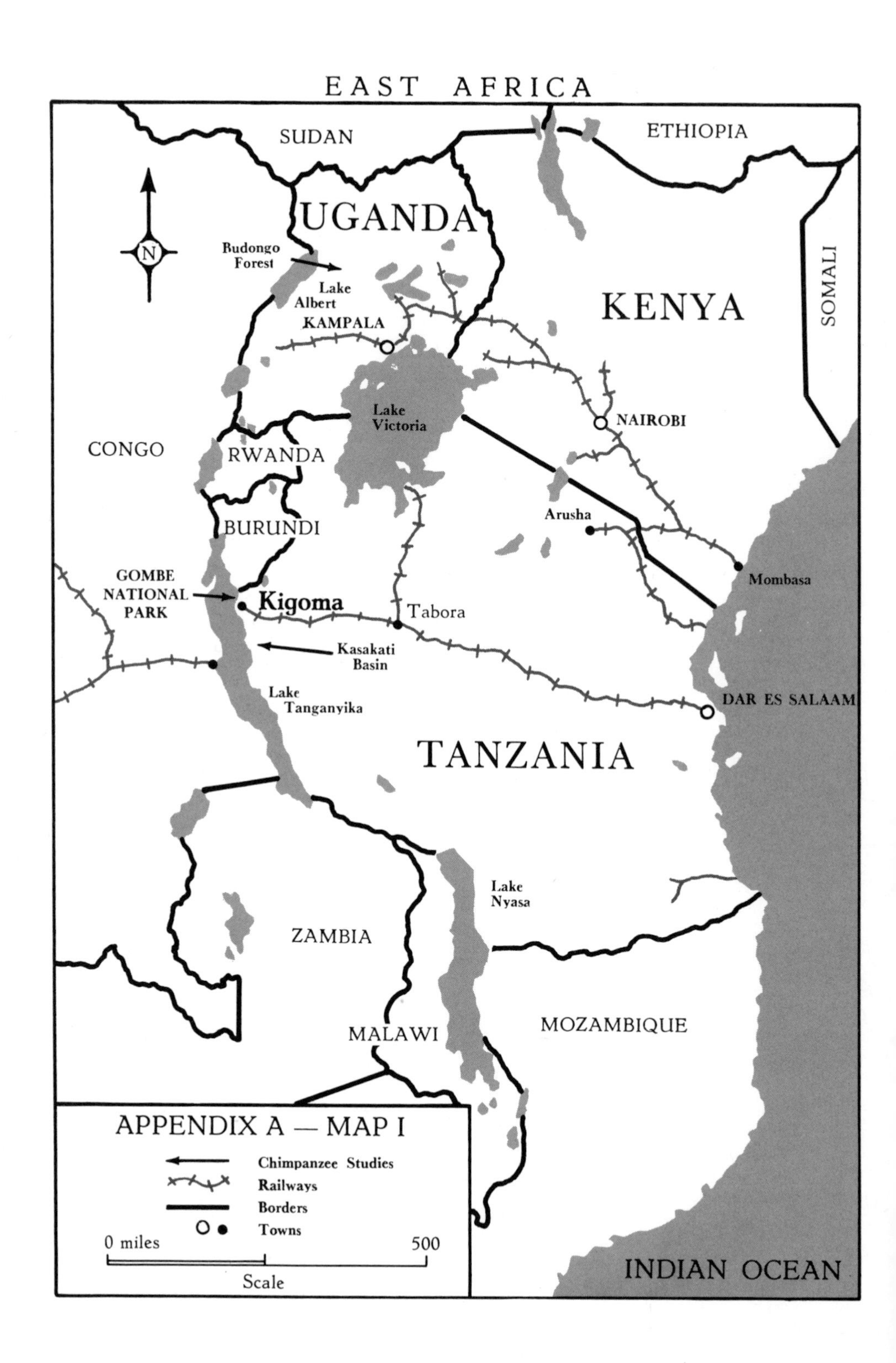
EAST AFRICA
SUDAN
ETHIOPIA
UGANDA
N
Budongo Forest
Lake Albert
KAMPALA
KENYA
SOMALI
Lake Victoria
NAIROBI
CONGO
RWANDA
BURUNDI
Arusha
Mombasa
GOMBE NATIONAL PARK
Kigoma
Tabora
Kasakati Basin
Lake Tanganyika
DAR ES SALAAM
TANZANIA
Lake Nyasa
ZAMBIA
MALAWI
MOZAMBIQUE
APPENDIX A — MAP I
Chimpanzee Studies
Railways
Borders
Towns
0 miles
500
Scale
INDIAN OCEAN

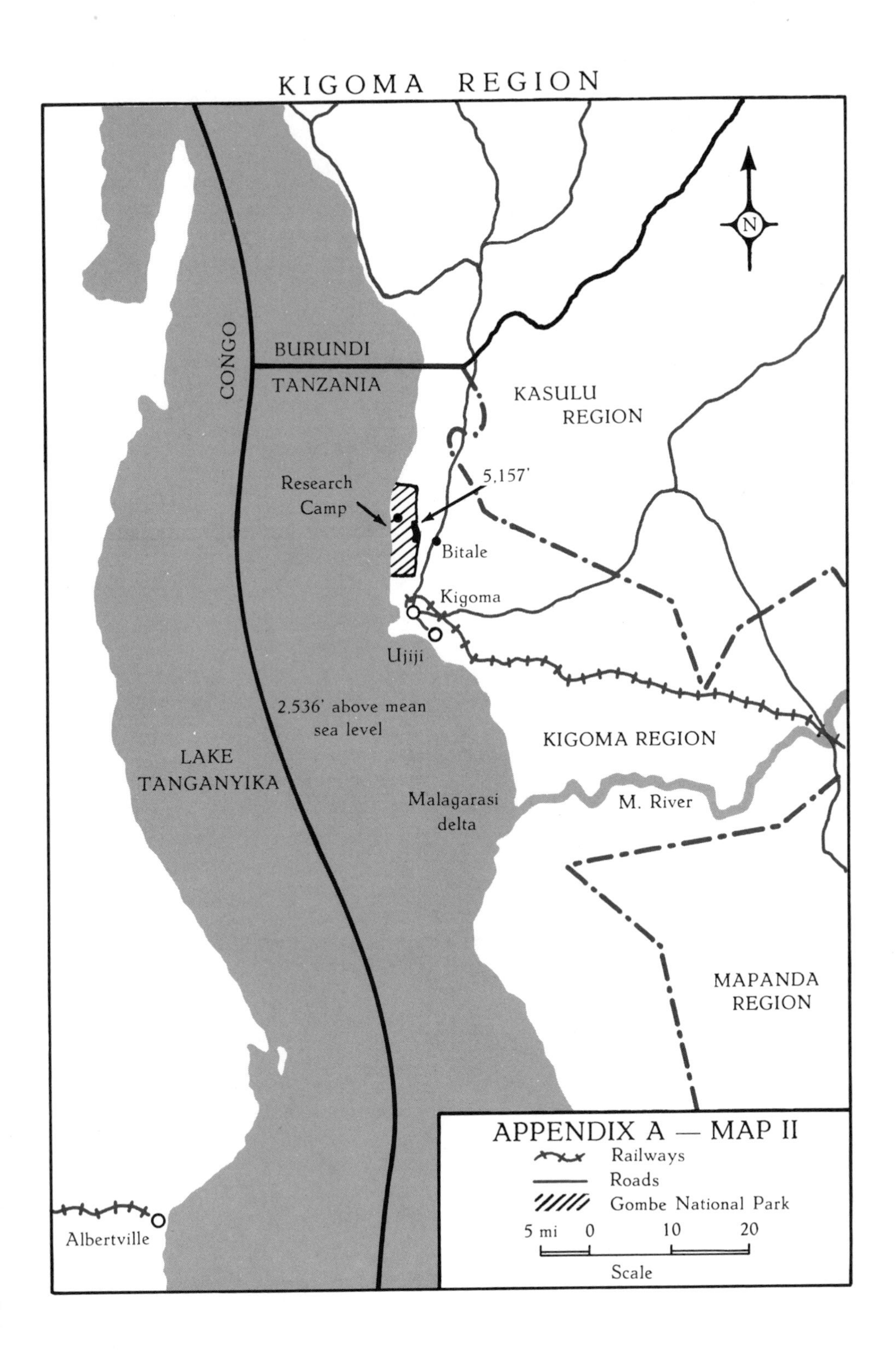
KIGOMA REGION
CONGO
BURUNDI
TANZANIA
KASULU
REGION
N
Research
Camp
5,157'
Bitale
Kigoma
Ujiji
2,536' above mean
sea level
LAKE
TANGANYIKA
KIGOMA REGION
Malagarasi
delta
M. River
MAPANDA
REGION
Albertville
APPENDIX A — MAP II
Railways
Roads
Gombe National Park
5 mi
0
10
20
Scale

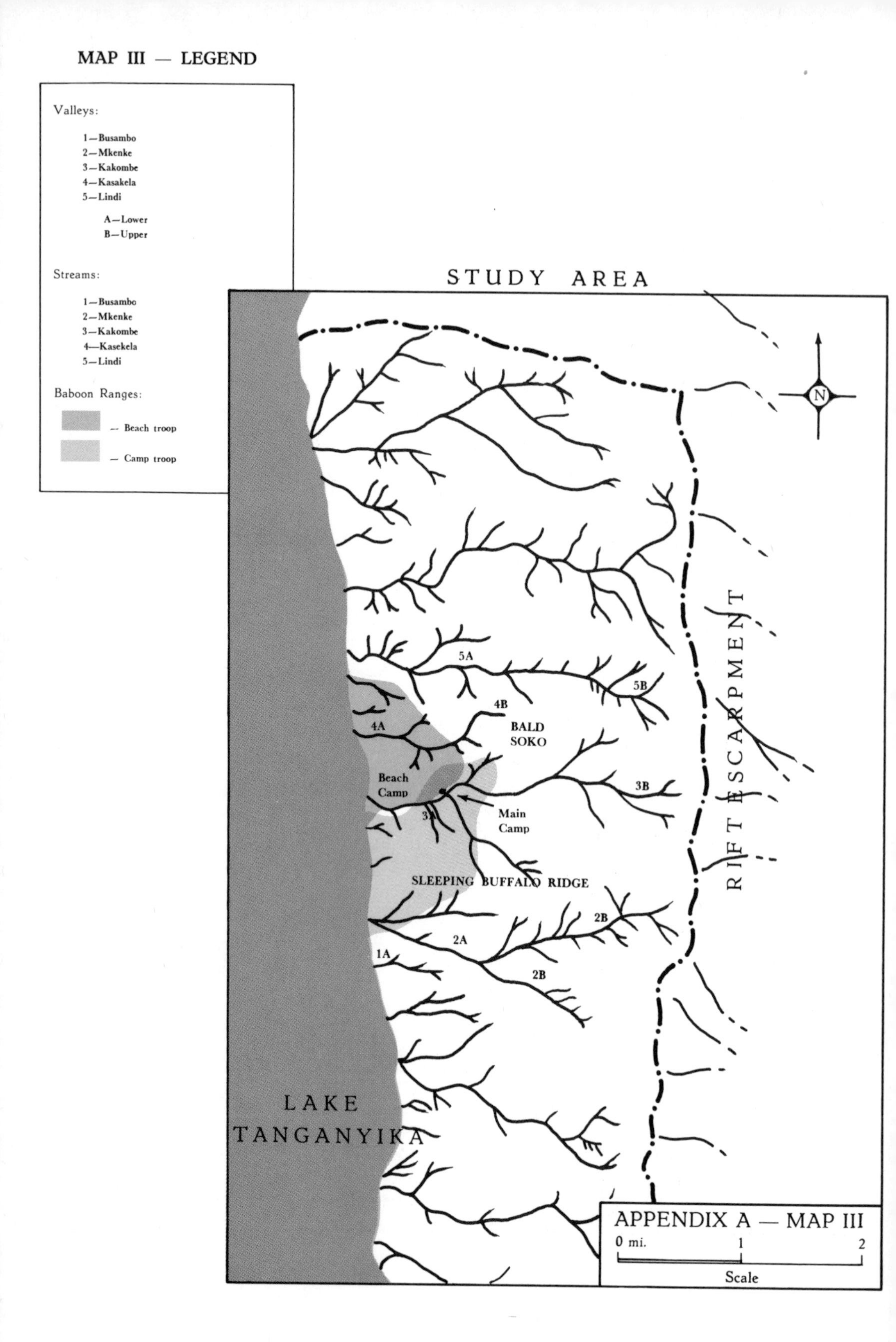

APPENDIX A — MAP III

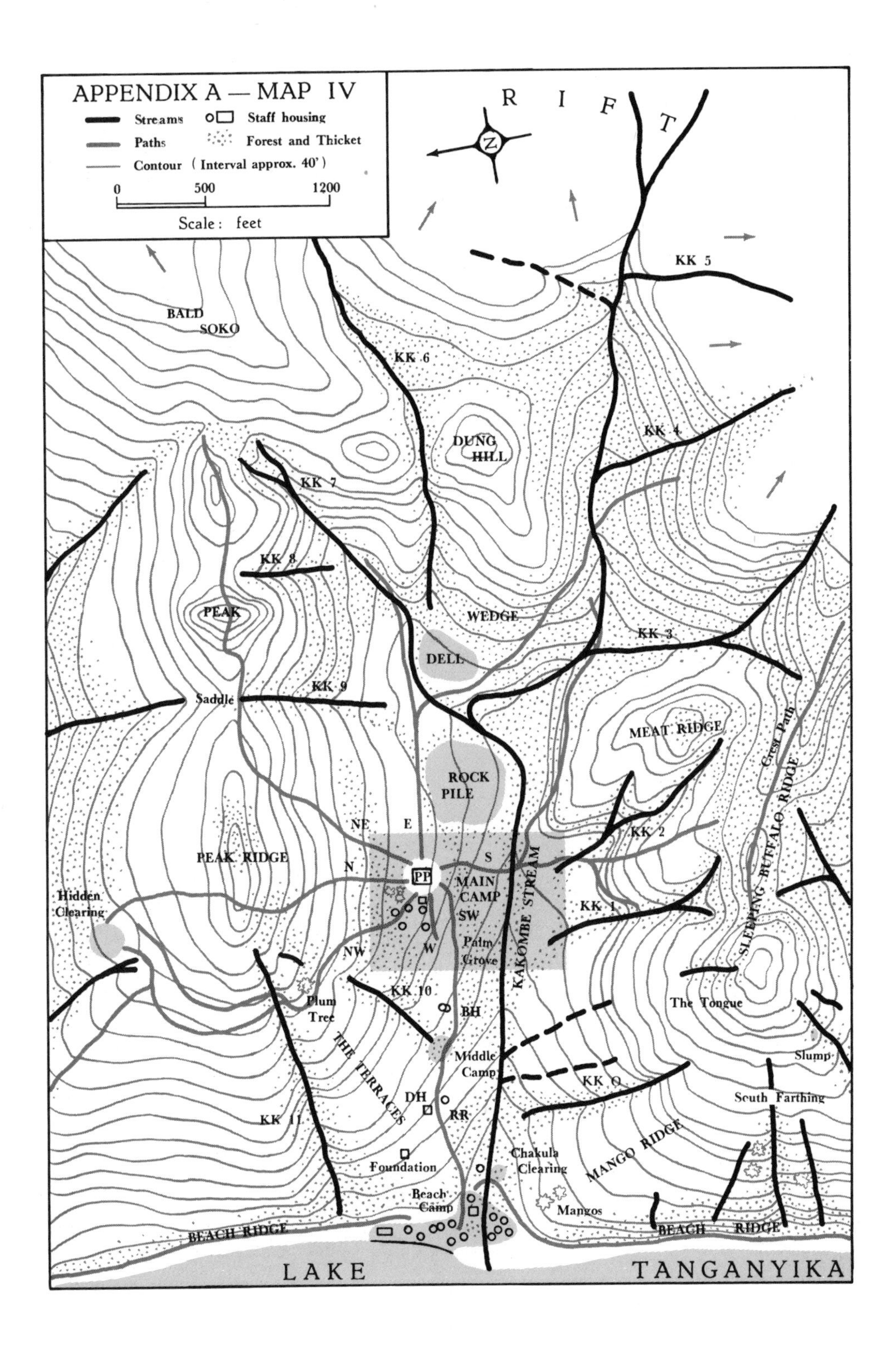
APPENDIX A — MAP IV
Streams
Staff housing
Paths
Forest and Thicket
Contour (Interval approx. 40')
0
500
1200
Scale : feet
R I F T
Z
BALD SOKO
KK 5
KK 6
DUNG HILL
KK 4
KK 7
KK 8
PEAK
WEDGE
KK 3
DELL
KK 9
Saddle
MEAT RIDGE
ROCK PILE
Crest Path
SLEEPING BUFFALO RIDGE
NE
E
KK 2
PEAK RIDGE
N
S
PP
MAIN CAMP
KAKOMBE STREAM
Hidden Clearing
SW
KK 1
W
NW
Palm Grove
KK 10
Plum Tree
BH
The Tongue
THE TERRACES
Middle Camp
Slump
KK O
DH
RR
South Farthing
KK 11
Chakula Clearing
MANGO RIDGE
Foundation
Beach Camp
Mangos
BEACH RIDGE
BEACH RIDGE
LAKE TANGANYIKA

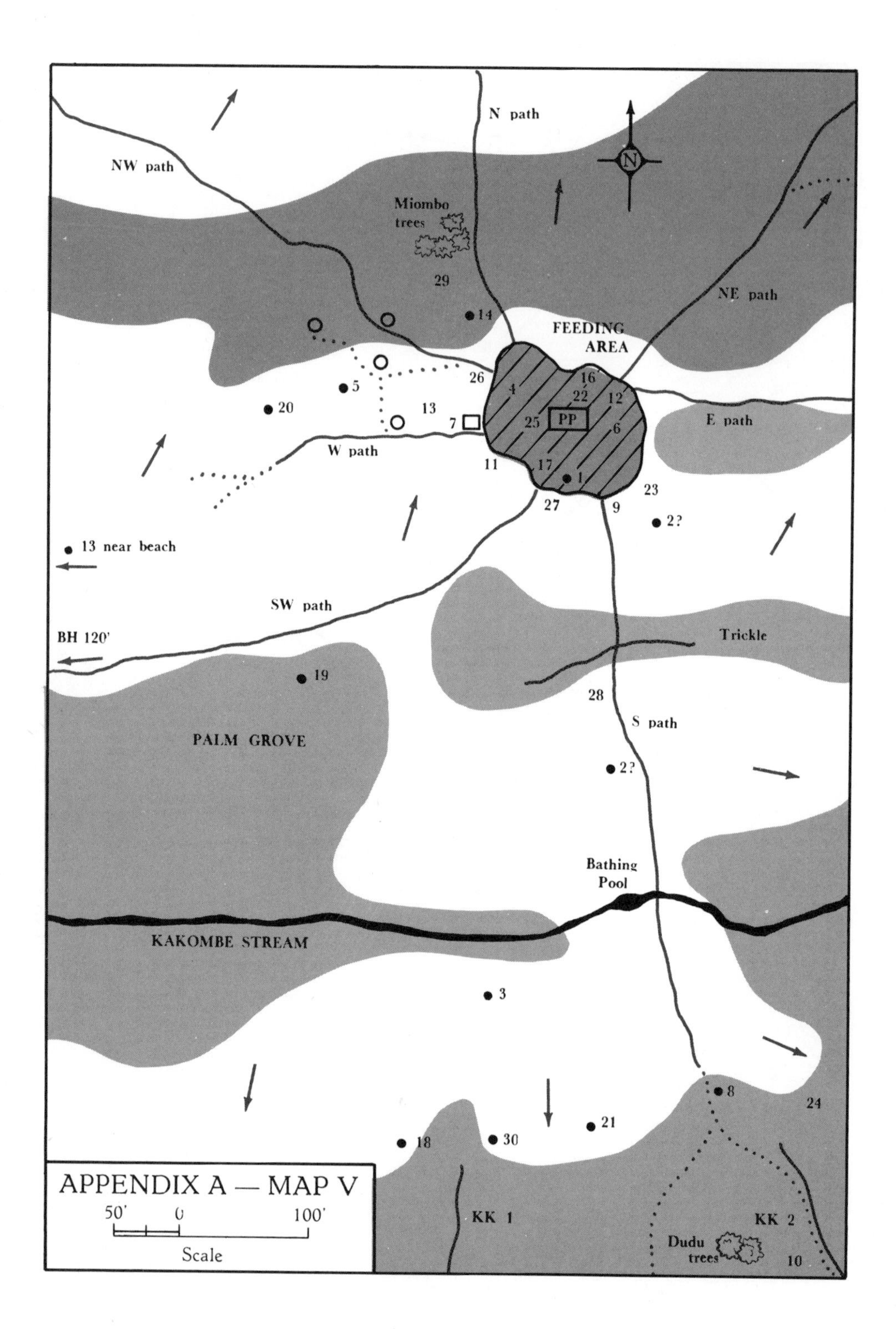

APPENDIX A — MAP V

MAP V — LEGEND

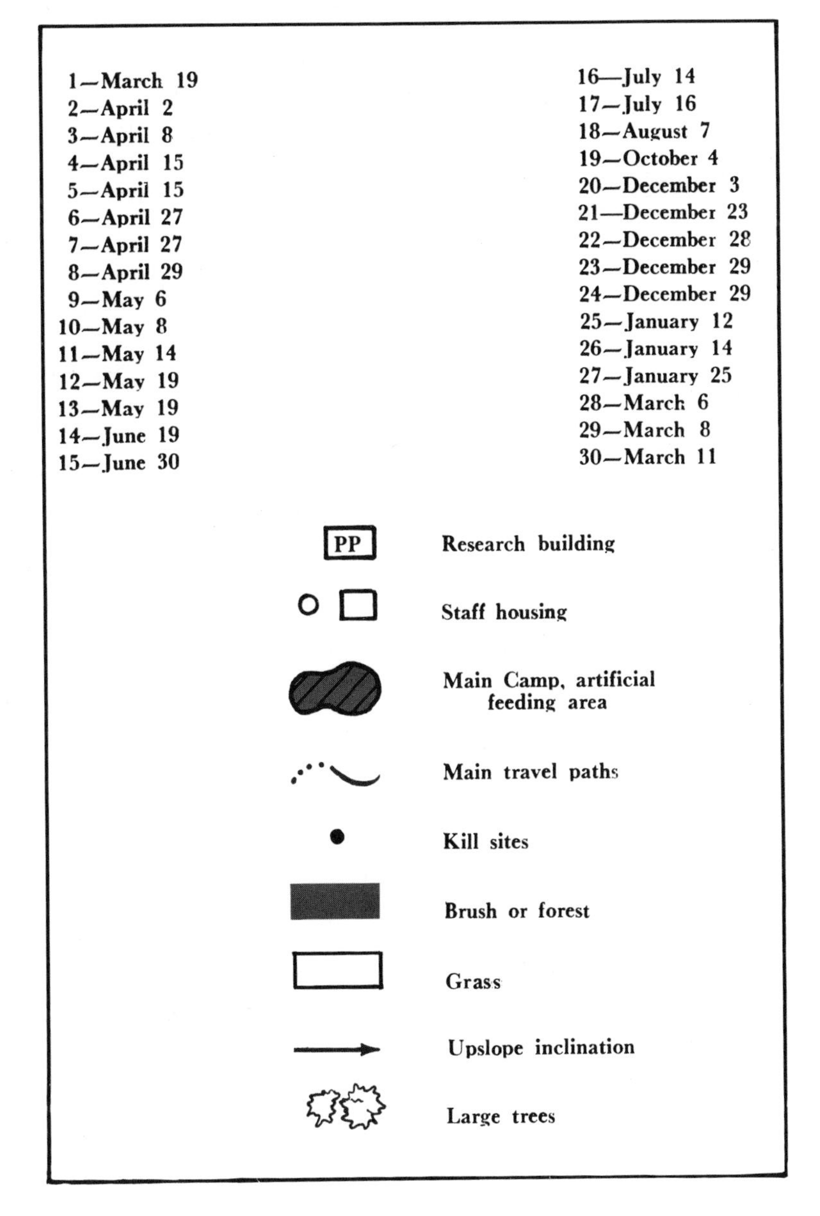
1—March 19
2—April 2
3—April 8
4—April 15
5—April 15
6—April 27
7—April 27
8—April 29
9—May 6
10—May 8
11—May 14
12—May 19
13—May 19
14—June 19
15—June 30
16—July 14
17—July 16
18—August 7
19—October 4
20—December 3
21—December 23
22—December 28
23—December 29
24—December 29
25—January 12
26—January 14
27—January 25
28—March 6
29—March 8
30—March 11
PP
Research building
Staff housing
Main Camp, artificial feeding area
Main travel paths
Kill sites
Brush or forest
Grass
Upslope inclination
Large trees

B. TABLES

Tables I and II provide individual background data on the chimpanzee study community. Table III gives information on the 30 episodes observed in 1968–69. Tables IV through VIII contain individual and numerical data on predatory behavior.

TABLE I

The Chimpanzee Study Community:
potential participants in predation
during the 1968-69 research period.

MALES:		*BORN:*
13 adults (Table II)		
Willy Wally	WW	1953 (paralytic)
De	DE	1956
Godi	GI	1957
Jomeo	JJ	1957 ± 1
Satan	ST	*1959*
Sniff	SF	*1961*
Sherry (Vodka*)	SH	*1962*
Flint (Flo)	FT	*1964* March 1
Goblin (Melissa)	GB	*1964* Sept 7
Mustard (Nope)	MU	*1965* Oct 29-Nov 22
Atlas (Athena)	AL	*1967* Sept 19-Oct 2
FEMALES:		
Flo	FLO	unknown (past prime)
Olly*	OL	unknown (past prime)
Passion	PS	unknown (prime)
Melissa	ML	1950 ± 3 ″
Mandy	MD	1950 ± 3 ″
Athena	AT	1954 ± 2 ″
Sprout	SP	unknown ″
Nope	NP	unknown ″
Madam Bee	MB	unknown ″ (paralytic)
Nova	NV	1954 ± 2 ″
Pallas	PL	1955 ± 2 ″
Gigi	GG	1956 ″
Miff	MF	*1958*
Fifi (Flo)	FF	*1959*
Winkle	WK	1959
Gilka (Olly)	GK	*1961*
Little Bee (M. Bee)	LB	*1961* (hered. or congen. defect.)
Sparrow (Sprout)	SW	1963-1964
Honey Bee (M. Bee)	HB	1964-1965
Cindy* (Circe*)	CD	*1965* Jan 22-24
Pom (Passion)	PM	*1965* July 13
Midge (Mandy)	MG	*1966* May 28-June 8
Flame* (Flo)	FL	*1968* Aug 25
Mooza (Miff)	MZ	*1969* Jan 20-22

* dead by early 1969
(mothers)
known dates

TABLE II

Adult Male Chimpanzees:
individuals most frequently involved in predation.

NAME:		*BORN:*	*ESTIMATED SOCIAL STATUS:*
Mike	MK	unknown (prime adult)	top ranking
— —			
Goliath	GOL	unknown (past prime)	high ranking
Leakey	LK	unknown "	"
Hugo	HG	unknown "	"
— —			
Hugh	HH	unknown (prime adult)	"
Humphrey	HM	1948 "	"
Rix [a]	RX	unknown "	(rank undetermined)
— —			
Worzle [b]	WZ	unknown (diseased)	low ranking
Faben (Flo)	FB	1953 (paralytic)	"
Pepe [c]	PP	1953 (paralytic)	"
— —			
Charlie	CH	1953 (young adult)	transitional
Evered (Olly)	EV	1955 "	"
Figan (Flo)	FG	1956 "	"

[a] died Nov 1968 [b] died April 1969 [c] disappeared Oct 1968
Note: the above table excludes WW and DE on grounds other than adulthood.

TABLE III

Descriptive Data on Predatory Episodes Observed in 1968-69.

Date	Time of Day	Duration[a]	Location[b]	Event	Mode[c]	Prey: Kind[d]	Prey: Identity	Prey: Age and Sex	Observers
March 19	8:20	312	1	Kill	1	B-1	*Amber*	4 weeks, ♀	CG, AS, TR, GT
April 2	9:24	270	2	Kill	2?	B-2	??	juvenile, ?	AS, GT
April 8	9:40	(207)	3	Kill	2	B-1	??	(juv.), ♀	CG, TR, BR, GT
April 15	12:28*	(12)	4	Attempt	1	B-2	Lane	16–24 weeks, ♂	CG, TR, PMcG, GT
	12:41*	149	5	Kill	2	B-2	*Beau*	40 weeks, ♂	CG, TR, PMcG, GT
April 27	12:32*	1	6	Attempt	?	B-1	??	(juv.), ?	CG, GT
	2:18*	16	7	Attempt	3	B-1	Thor	1 week, ♂	CG, GT
April 29	8:30	(105)	8	Kill	2?	B-2	*Dillon*	18 weeks, ♂	PMcG, GT
May 6	7:31	6	9	Attempt	2	B-?	??	??	GT
May 8	11:40	(16)	10	Attempt	?	B-?	??	??	PMcG
May 14	9:40	71	11	Attempt	3	B-1	Lamb	2 weeks, ♀	CG, BR, GT
May 19	7:12	10	12	Attempt	2	B-?	??	??	CG, GT
	11:05	12	13	Attempt	3	B-1	Lamb	3 weeks, ♀	CG, GT
June 19	11:08	127	14	Kill	1	B-1	*Thor*	8 weeks, ♂	CG, RD, PMcG, GT
June 30	11:30	(180)	15	Kill	?	B-2	*Huxley*	12 weeks, ♂	JvL, TR
July 14	7:34	5	16	Attempt	1	B-?	??	??	CG, RD
July 16	7:06	1	17	Attempt	1	B-1	Lamb	10 weeks, ♀	RD
Aug 7	8:10	540	18	Kill	?	C		adult ♀ or subad. ♂	CG, RD, PMcG
Oct 4	8:55	220	19	Kill	1	B-2	*Lane*	50 weeks, ♂	RD, DS, TR, AS, GT
Dec 3	11:26	(164)	20	Kill	2	B-1	*Soul*	12 weeks, ♀	PMcG, JG, RD, GT
Dec 23	12:31*	304	21	Kill	2	B-1	*Glad*	75 weeks, ♀	DS, PMcG, JG, GT
Dec 28	9:04	21	22	Attempt	1	B-1	Randy	32 weeks, ♀	RD, DS, GT
Dec 29	10:55	2	23	Attempt	2	B-1	??	(bl. inf.), ?	GT
	2:28*	2	24	Attempt	?	B-1	??	??	GT
Jan 12	9:04	8	25	Attempt	2?	B-1	??	(bl. inf.), ?	DS, RD, GT
Jan 14	11:00	2	26	Attempt	?	B-1	??	(trans. inf.), ?	RD, GT
Jan 25	10:31	3	27	Attempt	?	B-1	??	??	GT
March 6	1:55*	(15)	28	Attempt	2	B-?	??	??	CM
March 8	3:58*	(12)	29	Attempt	3?	B-?	??	??	CM
March 11	1:25*	(155)	30	Kill	?	bb		(infant), ?	RD

[a] minutes (obser. time only)
[b] see Map V
[c] 1—seizure, 2—chase, 3—stalk
[d] B—baboon (1-Camp, 2-Beach), C—colobus, bb—bushbuck

TABLE IV

Individual Data on Predation

Attempts	Potentially Participant Individuals	Active Participants	Stage I (Pursuit)
April 15	MK, HG, HH, HM, RX, WZ, FB, PP, CH, EV, FG, WW, gi, jj, st, sf/Ps-pm, Ml-gb, Md-mg, Np-mu, Gg, Mf, Ff, gk	MK, HG, RX, CH	*MK, HG,* RX, CH
April 27	MK, LK, HG, HH, HM, RX, WZ, FB, PP, CH, EV, FG, gi, jj, st, sf/Ps-pm, Ml-gb, Md-mg, At-al, Np-mu, Gg	MK, HH, RX, WZ, CH, FG	RX, WZ, *FG*
April 27	MK, LK, HG, HH, HM, RX, WZ, FB, PP, CH, EV, FG, gi, jj, st, sf/Ps-pm, Ml-gb, Md-mg, At-al, Gg, Mf, Ff	MK, HG, HH, HM, CH	MK, HG, HH, HM, CH
May 6	MK, LK, HM, RX, WZ, PP, CH, WW, jj, st, sf/Flo-ft, Ol, Ps-pm, Ml-gb, Np-mu, gk	MK, LK, HM, RX, WZ, CH	*MK,* LK, HM, RX, CH
May 8	MK, GOL, LK, HG, HH, HM, RX, PP, CH, EV, FG, WW, DE, gi, jj, st, sf/Flo-ft, Ol, Ps-pm, Ml-gb, Md-mg, At-al, Np-mu, Gg, Mf, Ff, gk	MK, GOL, LK, HG, HH, RX, CH	??
May 14	MK, GOL, LK, HG, HH, HM, RX, WZ, PP, CH, EV, FG, WW, jj, st, sf/Flo-ft, Ol, Ps-pm, Ml-gb, Md-mg, At-al, Np-mu, Gg, Ff, gk	MK, FG	MK, *FG*
May 19	RX, CH, FG, jj, st/Ps-pm, At-al, Nv	CH, FG	*CH*
May 19	MK, GOL, LK, HG, HH, HM, RX, PP, CH, FG, WW, gi, jj, st, sf/Ol, Ps-pm, Md-mg, At-al, Np-mu, Nv, Gg, Mf, Ff, gk	HG, FG	HG, *FG*
July 14	MK, HG, HH, WZ, CH, sf/Flo-ft, Ml-gb, Nv, Pl	MK, HG, CH	*MK,* HG, CH
July 16	MK, HH, WZ, PP, FG	MK	*MK*
Dec 28	MK, HG, HH, FB, st, ft/Flo-fm, Ps-pm, Gg, wk	MK, HG	*MK, HG*
Dec 29	MK, HG, HM, FB, CH, EV, FG, WW, DE, jj, st, sf, ft/Flo-fm, Ps-pm, Ml-gb, Md-mg, At-al, Np-mu, Nv, Pl, Mf, wk, gk	HG, HM, CH	*HG,* HM, *CH*
Dec 29	HG, HM, CH, FB, WW, ft/Flo-fm	HG, HM, CH	??
Jan 12	MK, HM, FB, EV, jj, st, ft/Flo-fm, Ol, Ps-pm, At-al, Nv, Pl, Gg, Mf, Ff, gk	MK	*MK*
Jan 14	MK, LK, HG, sf	MK, LK, HG	*MK,* LK, *HG*
Jan 25	MK, WZ, FB, jj, st, ft/Flo-fm, Ps-pm, Np-mu, Gg, Mf-mz, Ff, gk	MK	*MK*
March 6	MK, GOL, LK, HG, HH, HM, WZ, FG, jj, st, sf/Ml-gb, At-al, Mf-mz, wk, gk	MK, HG, HH, HM	??
March 8	MK, GOL, LK, HG, HH, WZ, FB, EV, FG, jj, st, sf/Flo-ft, Ps-pm, Ml-gb, At-al, Np-mu, Mf-mz, Ff, gk	MK, GOL, HG, HH, st, sf	*MK, GOL,* HG, HH

XX—adult male
Xx—adult female
xx—adolescent-infant

XX—instigator
Xx-xx—mother-infant
←males/females→

TABLE V – Individual Data on Predation

Kills	Potentially Participant Individuals	Total Number of Participants	Active Participants Stage I (Pursuit)	Active Participants Stage II (Capture)
March 19	MK, GOL, LK, HH, HM, RX, WZ, PP, CH, EV, FG, jj, st, sf/Flo-ft, Md-mg, At-al, Np-mu, Nv, Pl, Ff, gk	MK, GOL, LK, HH, HM, RX, WZ	*MK,* LK, HH, RX	HM?
April 2	MK, GOL, HG, HH, HM, RX, WZ, FB, PP, CH, EV, FG, jj, st, sf/Flo-ft, Ps-pm, Ml-gb, Md-mg, At-al, Np-mu, Gg, Ff	MK, GOL, HG, HH, HM, RX, WZ, FB, CH, EV, sf/Flo-ft, Md (-mg), Np-(-mu), Ff	MK, *RX?*	RX?, MK?
April 8	MK, GOL, LK, HG, HM, RX, WZ, FB, PP, CH, EV, FG, jj, st, sf/Flo-ft, Ol, Ps-pm, Ml-gb, Md-mg, At-al, Np-mu, Gg, Ff, gk	MK, GOL, LK, HG, HM, WZ, PP, sf/Flo-ft, Ol, Ff, gk	MK?, GOL?, LK?, HG?, HM?	MK, GOL, LK, HG
April 15	MK, HG, HH, HM, WZ, FB, CH, EV, FG, WW, gi, jj, st, sf/Ps-pm, Ml-gb, At-al, Sp-sw, Np-mu, Nv, Gg, Mf, Ff, gk	MK, HG, RX, CH/Ml (-gb), Sp (-sw), Mf, Ff	MK, *HG,* RX, CH	HG
April 29	MK, GOL, LK, HG, HH, HM, RX, WZ, PP, CH, EV, FG, st, sf/Flo-ft, Ps-pm, Ml-gb, Md-mg, At-al, Np-mu, Nv, Pl, Gg, Mf, Ff	MK, GOL, LK, CH, EV/Ff	LK, *CH*	LK, CH
June 19	MK, GOL, LK, HG, HH, HM, RX, WZ, CH, EV, FG, WW, DE, gi, jj, st, sf/Flo-ft, Ol, Ps-pm, Ml-gb, At-al, Np-mu, Nv, Pl, Gg, Mf, Ff, gk	MK, GOL, LK, HG, HH, HM, CH/Ol, Nv, Gg	*MK,* HG, *HH,* CH	MK, HG, HH, CH
June 30	GOL, WZ, jj/Flo-ft	GOL, WZ, jj/Flo-ft	*GOL,* WZ?	GOL
Aug 7	MK, GOL, WZ, FB, CH, EV, FG, WW, DE, gi, jj, st, sf, sh/Flo-ft, Ps-pm, Ml-gb, Md-mg, At-al, Np-mu, Mf, Ff	MK, GOL, FB, CH, EV, FG, st, sf, sh/Flo-ft, Ml (-gb), Md(-mg), At-al, Np (-mu), Ff	GOL, *FG?*	GOL?
Oct 4	MK, LK, HG, HH, RX, FB, CH, FG, WW, gi, jj, st, sf/Ps-pm, Ml-gb, Md-mg, Pl, Gg, wk	MK, LK, HG, HH, RX, FB, CH, WW, gi, jj, sf, gb/Gg, pm	MK, HG, *FB*	MK, HG
Dec 3	MK, GOL, LK, HG, HH, HM, WZ, FB, CH, WW, gi, jj, st, sf, ft/Flo-fm, Ol, Ps-pm, Ml-gb, Md-mg, At-al, Np-mu, Pl, Gg, Mf, Ff, wk, gk	MK, GOL, HG, HH, HM, CH, st, sf, ft/Flo (-fm), Gg, Ff, gk	MK, GOL, *HG,* HH, HM, CH	HG, HM?
Dec 23	MK, GOL, LK, HH, HM, WZ, CH, EV, FG, DE, gi, jj, st, sf, ft/Flo-fm, Ps-pm, At-al, Np-mu, Nv, Pl, Mf, Ff, wk, gk	MK, GOL, LK, HH, HM, WZ, CH, FG, st, sf, ft/Flo (-fm), Nv, Ff	MK, GOL, HH, HM, *CH,* FG, sf	MK, GOL, HH, CH
March 11	LK, HG, CH, FG, gi, jj, sf/Mb-hb, Pl, Mf-mz, wk, lb	LK, HG, CH, FG, gi, jj, sf/Mb-hb, Pl, wk, lb	*LK, HG,* CH, FG	LK?, HG?

XX–adult male, Xx–adult female, xx–adolescent-infant;

TABLE V (Cont'd.) – Individual Data on Predation

Active Participants				Interactive Participants	
Stage III (Consumption)					
Total Number of Consumers	Consumers of Major Portions	Consumers of Brain	Collectors of Pieces	Begging By ( ) From ( )	Rewarded Begging
MK, GOL, HM, WZ	GOL, HM	??	??	MK-GOL, LK-GOL	MK
MK, GOL, HG, HH, WZ, CH, EV, sf/Flo-ft, Md, Ff	MK, GOL, HG	MK	sf, ft	Ff-MK, Np-MK	Ff
MK, GOL, LK, HG, HM, WZ, PP, sf/Flo-ft, Ol, Ff, gk	MK, GOL, LK, HG, HM	MK	WZ, PP, sf, ft/Ff	Ol-MK, PP-LK, ft-Flo, gk-Ol, ft-MK	Ol, PP, gk
MK, HG/Ml, Sp	HG	HG	??	MK-HG, Ml-HG, Sp-HG	MK, Ml, Sp
MK, LK, CH, EV/Ff	LK, CH	LK	??	GOL-LK, MK-LK, Ff-LK	GOL, MK
MK, HH, CH/Nv, gk	MK, HH	MK	??	CH-HH, Ol-MK, Nv-MK, gk-HH	CH, Nv
GOL, WZ, jj/Flo-ft	GOL, WZ/Flo	GOL	jj	WZ-GOL, Flo-GOL, ft-Flo, ft-GOL	WZ, Flo, ft
MK, GOL, FB, CH, EV, FG, sf, sh/Flo-ft, At-al, Md, Np, Ff	MK, GOL	MK	sf, sh/At	Flo-MK, GOL-MK, ft-Flo, Ml-MK, CH-MK, Ff-MK, Flo-Ff, ft-Ff, FB-MK, At-Flo, At-Ff, al-Ff, CH-FB, At-FB, Flo-FB, al-At, FG-GOL, Md-GOL, Np-MK, ft-Md, st-FG	Flo, GOL, ft, Ff al, FG, Md, Np
MK, LK, HG, HH, WW, jj, sf/Gg	MK, HH, CH	MK	WW, jj, sf/Gg	CH-MK, Gg-MK, Gg-HG, LK-MK, pm-WW	Gg, LK
MK, GOL, HG, st, ft/Flo	MK, HG	HM?	st, ft	MK-HG, GOL-HG, Gg-MK, Flo-HG	MK, GOL
MK, GOL, LK, HM, WZ, CH, FG, st, sf, ft/Flo, Ff	MK, GOL	MK	HM, st, sf/Flo	LK-MK, HM-MK, Flo-MK, Ff-MK, Nv-MK, HM-LK, WZ-GOL, Nv-GOL, HM-GOL, Flo-GOL	LK, Ff, HM, WZ, HM
LK, HG, CH, FG, gi, sf/Mb, wk, lb	LK, HG, FG	HG	FG, gi, sf/wk	FG-HG, sf-HG, Mb-HG, wk-Mb, hb-Mb, lb-Mb, hb-FG, CH-LK, wk-LK	FG, sf, Mb, lb, CH

←males/females→; *XX*–captor; Xx-xx–mother-infant

TABLE VI Numerical Data on Predation

	EVENTS	Total # of Potential Participants				Total # of Participants				Participants in Stage I (Pursuit)				Participants in Stage II (Capture)				Participants in Stage III (Consumption)			
			XX	Xx	xx		XX	Xx	xx		XX	Xx	xx		XX	Xx	xx		XX	Xx	xx
Kills	March 19	26	11	7	8	7	7	0	0	4	4	0	0	?	?	?	?	4	4	0	0
	April 2	29	12	8	9	16	10	4	2	2	2	0	0	2	2	0	0	12	7	3	2
	April 8	31	12	9	10	13	7	3	3	5	5	0	0	4	4	0	0	13	7	3	3
	April 15	29	10	9	10	8	4	4	0	4	4	0	0	1	1	0	0	4	2	2	0
	April 29	31	12	11	8	6	5	1	0	2	2	0	0	2	2	0	0	5	4	1	0
	June 19	34	13	11	10	10	7	3	0	4	4	0	0	4	4	0	0	5	3	1	1
	June 30	5	2	1	2	5	2	1	2	2	2	0	0	1	1	0	0	5	2	1	2
	Aug 7	28	9	8	11	17	6	6	5	2	2	0	0	?	?	?	?	15	6	5	4
	Oct 4	22	9	5	8	14	8	1	5	3	3	0	0	2	2	0	0	8	5	1	2
	Dec 3	34	10	11	13	13	6	3	4	6	6	0	0	2	2	0	0	6	3	1	2
	Dec 23	29	10	8	11	14	8	3	3	7	6	0	1	4	4	0	0	12	7	2	3
	March 11	14	4	3	7	12	4	2	6	4	4	0	0	2	2	0	0	9	4	1	4
	Sub-totals	312	114	91	107	135	74	31	30	45	44	0	1	24	24	0	0	98	54	21	23
Attempts	April 15	28	12	7	9	4	4	0	0	4	4	0	0	—	—	—	—	—	—	—	—
	April 27	27	12	6	9	6	6	0	0	3	3	0	0	—	—	—	—	—	—	—	—
	April 27	27	12	7	8	5	5	0	0	5	5	0	0	—	—	—	—	—	—	—	—
	May 6	21	8	5	8	6	6	0	0	6	6	0	0	—	—	—	—	—	—	—	—
	May 8	34	13	10	11	7	7	0	0	7	7	0	0	—	—	—	—	—	—	—	—
	May 14	32	13	9	10	2	2	0	0	2	2	0	0	—	—	—	—	—	—	—	—
	May 19	10	3	3	4	2	2	0	0	1	1	0	0	—	—	—	—	—	—	—	—
	May 19	29	11	9	9	2	2	0	0	2	2	0	0	—	—	—	—	—	—	—	—
	July 14	12	5	4	3	3	3	0	0	3	3	0	0	—	—	—	—	—	—	—	—
	July 16	5	5	0	0	1	1	0	0	1	1	0	0	—	—	—	—	—	—	—	—
	Dec 28	12	4	3	5	2	2	0	0	2	2	0	0	—	—	—	—	—	—	—	—
	Dec 29	30	9	9	12	3	3	0	0	3	3	0	0	—	—	—	—	—	—	—	—
	Dec 29	8	5	1	2	3	3	0	0	3	3	0	0	—	—	—	—	—	—	—	—
	Jan 12	20	4	9	7	1	1	0	0	1	1	0	0	—	—	—	—	—	—	—	—
	Jan 14	4	3	0	1	3	3	0	0	3	3	0	0	—	—	—	—	—	—	—	—
	Jan 25	17	3	6	8	1	1	0	0	1	1	0	0	—	—	—	—	—	—	—	—
	March 6	19	8	3	8	4	4	0	0	4	4	0	0	—	—	—	—	—	—	—	—
	March 8	26	9	7	10	6	4	0	2	4	4	0	0	—	—	—	—	—	—	—	—
	Sub-totals	361	139	98	124	61	59	0	2	55	55	0	0								
	TOTALS	673	253	189	231	196	133	31	32	100	99	0	1	(24)	(24)	(0)	(0)	(98)	(54)	(21)	(23)

XX—adult males Xx—adult females
xx—adolescent-infant

Note: These data compiled from Table IV and Table V.

Note: the chimpanzee study population consists of 50% males and 50% females, and of the total number 31% are adult males and 30% adult females.

Basic Computations:

1. $477/673 = 71\%$ non-participation (of total potential)
2. $196/673 = 29\%$ participation (of total potential)
   - A. $61/361 = 17\%$ participation in attempts
   - B. $135/312 = 43\%$ participation in kills
     - a. $100/673 = 15\%$ participation in Stage I (100% males)
     - b. $24/312 = 8\%$ participation in Stage II (100% males)
     - c. $98/312 = 25\%$ participation in Stage III (55% males)

TABLE VII

Predatory Participation and Carcass Division Data:
a possible indicator of relative social status among adult males.

	MK	GOL	HG	LK	HH	HM	CH	WZ	RX	FB	PP	EV	FG
March 19	x *	x * ●		x *	x *	x * 0	x	x *	x *	x *	x	x	x
April 2	x * ●	x * 0	x * 0		x *	x *	x *	x *	x *	x *	x	x *	x
April 8	x * ●	x * 0	x * 0	x * 0		x * 0	x	x *	x	x	x *	x	x
April 15	x *		x * ●		x	x	x *	x		x		x	x
April 29	x *	x *	x	x * ●	x	x	x * 0	x	x		x	x *	x
June 19	x * ●	x *	x *	x *	x * 0	x *	x *	x	x			x	x
June 30		x * ●						x * 0					
October 4	x * ●	x * 0	x *	x *	x * 0		x * 0		x *	x *			x
December 3	x * 0	x *	x * 0	x	x *	x *	x *	x		x			
December 23	x * ●	x * 0		x *	x *	x *	x *	x *				x	x *
Totals	9 9 6	9 9 6	7 6 4	7 6 2	8 6 2	8 6 2	9 7 2	9 5 1	6 3 0	6 3 0	4 1 0	7 2 0	8 1 0
Categories	1		2					3				4	
Participates / **Present** =	100%	100%	86%	86%	75%	75%	79%	55%	50%	50%	28%	25%	12%
Rewarded / **Present** =	67%	67%	57%	29%	25%	25%	22%	11%	—	—	—	—	—
Rewarded / **Participates** =	67%	67%	67%	34%	34%	34%	29%	20%	—	—	—	—	—

x — present at predatory episode

* — participates in predatory episode

0 — obtains a large portion

● — consumes the brain

TABLE VIII

Numerical Data on Meat Procurement Interactions during Consumption (Stage 3): Tabulated for the Most Frequent Participants of the Three Major Age-sex Classes.

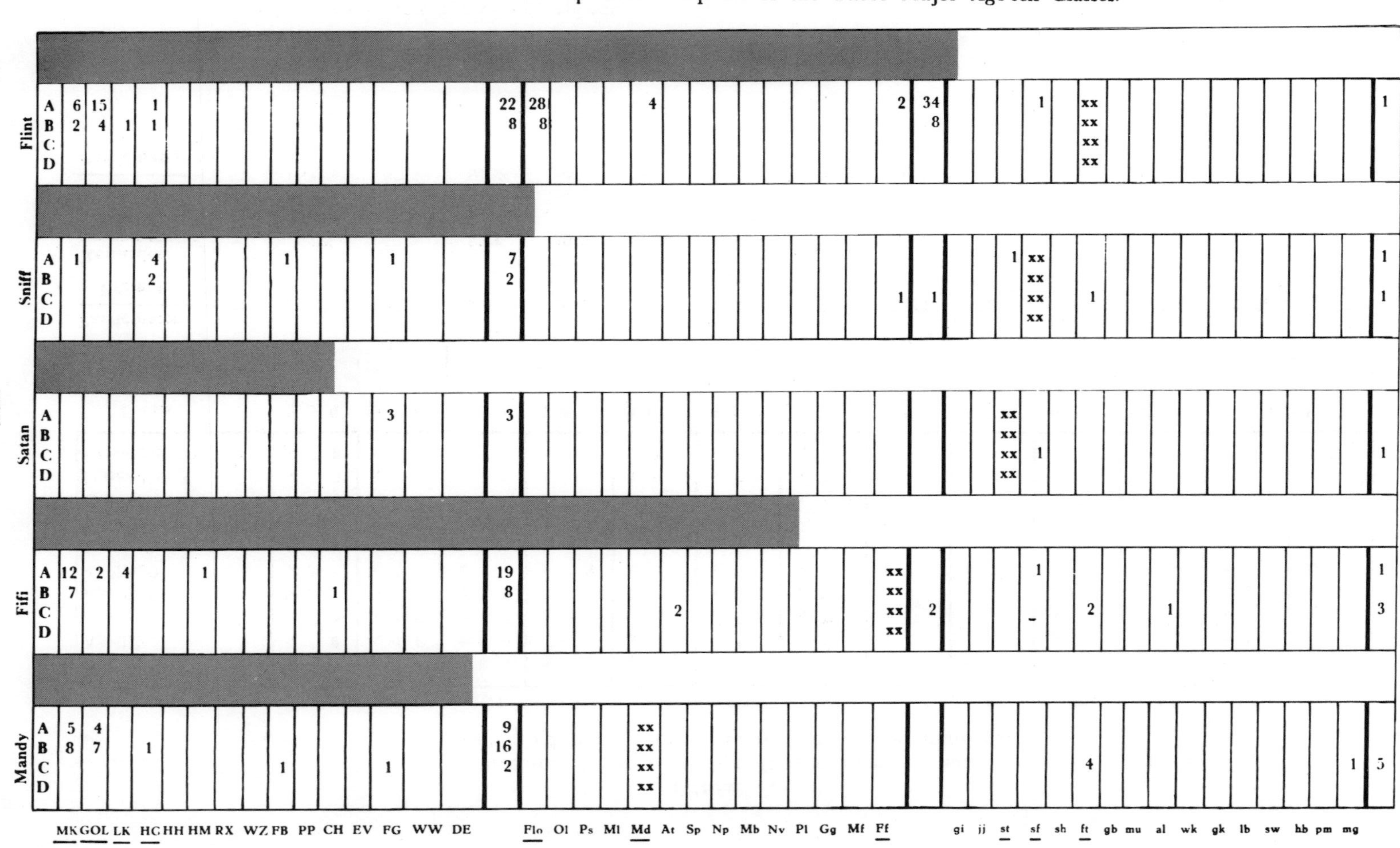

		MK	GOL	LK	HC	HH	HM	RX	WZ	FB	PP	CH	EV	FG	WW	DE	
Flint	A	6	15		1												22
	B	2	4	1	1												8
	C																
	D																
Sniff	A	1			4					1				1			7
	B				2												2
	C																
	D																
Satan	A													3			3
	B																
	C																
	D																
Fifi	A	12	2	4			1										19
	B	7										1					8
	C																
	D																
Mandy	A	5	4														9
	B	8	7		1												16
	C									1				1			2
	D																

		Flo	Ol	Ps	Ml	Md	At	Sp	Np	Mb	Nv	Pl	Gg	Mf	Ff	
Flint	A	28				4									2	34
	B	8														8
	C															
	D															
Sniff	A															
	B															
	C														1	1
	D															
Satan	A															
	B															
	C															
	D															
Fifi	A														xx	
	B														xx	
	C						2								xx	2
	D														xx	
Mandy	A					xx										
	B					xx										
	C					xx										
	D					xx										

		gi	jj	st	sf	sh	ft	gb	mu	al	wk	gk	lb	sw	hb	pm	mg	
Flint	A				1		xx											1
	B						xx											
	C						xx											
	D						xx											
Sniff	A			1	xx													1
	B				xx													
	C				xx		1											1
	D				xx													
Satan	A			xx														
	B			xx														
	C			xx	1													1
	D			xx														
Fifi	A				1													1
	B																	
	C				-		2			1								3
	D																	
Mandy	A																	
	B																	
	C						4										1	5
	D																	

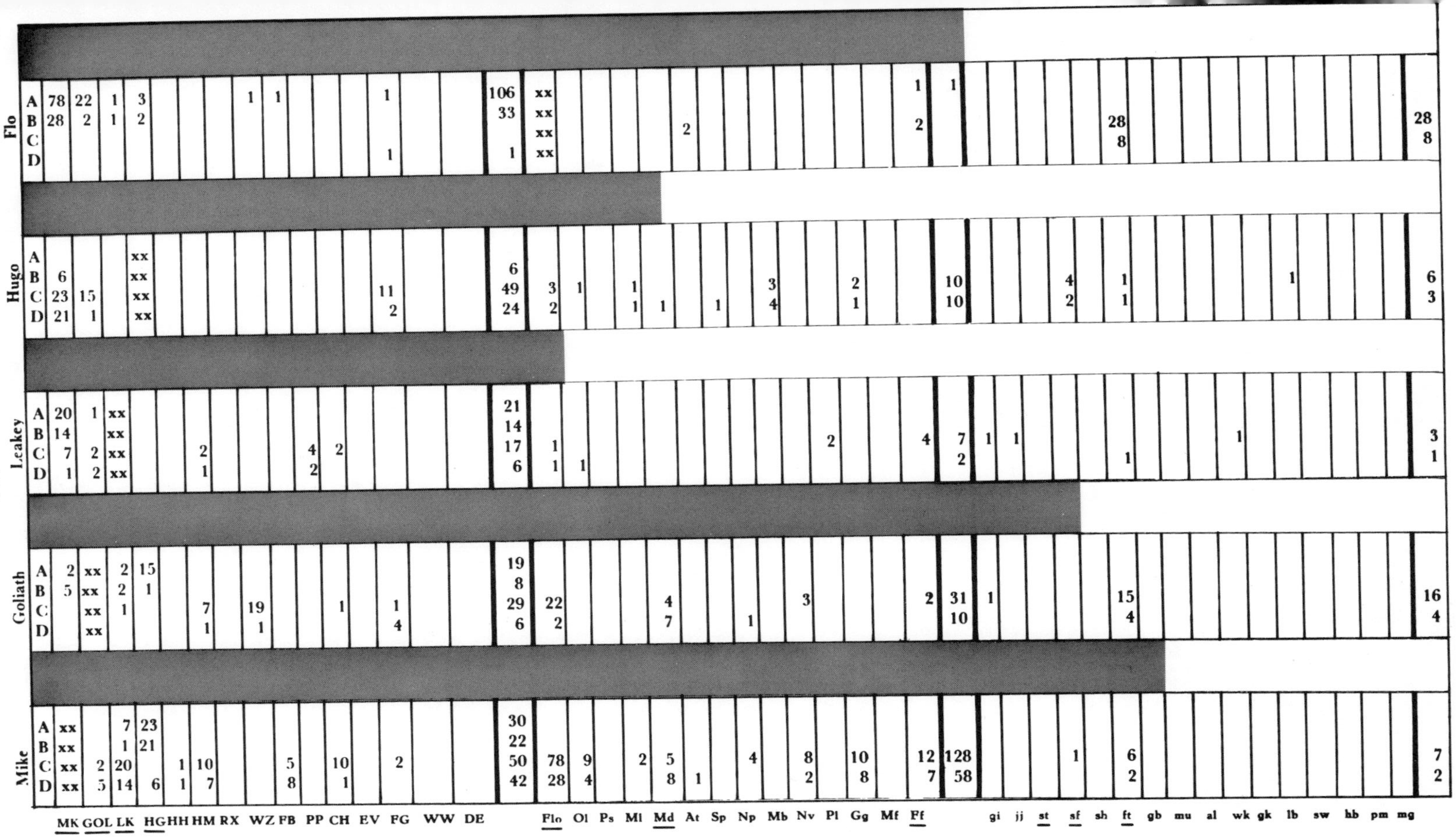

		MK	GOL	LK	HG	HH	HM	RX	WZ	FB	PP	CH	EV	FG	WW	DE	
Flo	A	78	22	1	3				1	1				1			106
	B	28	2	1	2												33
	C																
	D													1			1
Hugo	A				xx												
	B	6			xx												6
	C	23	15		xx									11			49
	D	21	1		xx									2			24
Leakey	A	20	1	xx													21
	B	14		xx													14
	C	7	2	xx			2				4	2					17
	D	1	2	xx			1				2						6
Goliath	A	2	xx	2	15												19
	B	5	xx	2	1												8
	C		xx	1			7		19			1		1			29
	D		xx				1		1					4			6
Mike	A	xx		7	23												30
	B	xx		1	21												22
	C	xx	2	20		1	10			5		10		2			50
	D	xx	5	14	6	1	7			8		1					42

		Flo	Ol	Ps	Ml	Md	At	Sp	Np	Mb	Nv	Pl	Gg	Mf	Ff	
Flo	A	xx													1	1
	B	xx														
	C	xx					2								2	
	D	xx														
Hugo	A															
	B															
	C	3	1		1					3			2			10
	D	2			1	1		1		4			1			10
Leakey	A															
	B															
	C	1										2			4	7
	D	1	1													2
Goliath	A															
	B															
	C	22				4					3				2	31
	D	2				7			1							10
Mike	A															
	B															
	C	78	9		2	5			4		8		10		12	128
	D	28	4			8	1				2		8		7	58

		gi	jj	st	sf	sh	ft	gb	mu	al	wk	gk	lb	sw	hb	pm	mg	
Flo	C						28											28
	D						8											8
Hugo	C				4		1						1					6
	D				2		1											3
Leakey	C	1	1								1							3
	D						1											1
Goliath	C	1					15											16
	D						4											4
Mike	C				1		6											7
	D						2											2

A — Begs from
B — Takes from
C — Begged from
D — Taken from

Attendance Time (1 hr = 1 cm) at Consumption

C. MEAT DISTRIBUTION DIAGRAMS

The following ten diagrams were selected to illustrate meat distribution. The events of March 19 and April 29 have been excluded due to lack of observational detail. Each diagram includes a main distribution pattern as well as a list (at the right) of those individuals who collected fragments from the ground. Each event is additionally broken down into half hour time periods for convenience, with double vertical lines connecting the periods.

The following key provides specific information on the manner of distribution:

MEAT DISTRIBUTION KEY

▬▬▬ (thick line)	initial division or later sharing of large portions
( )	possessors of major portions
□	consumption of brain
— — — -	meat touched but nothing taken
-●-●-●-●-●-	possessor of meat closely approached and watched
—•—•—	overt gestural or vocal request for meat, **unsuccessful**
....	overt gestural or vocal request for meat, **successful**
———	meat briefly chewed while in the possession of another
—➤—➤	meat voluntarily held toward or handed to another

DIAGRAM I

Stage 3 — April 2, 1968

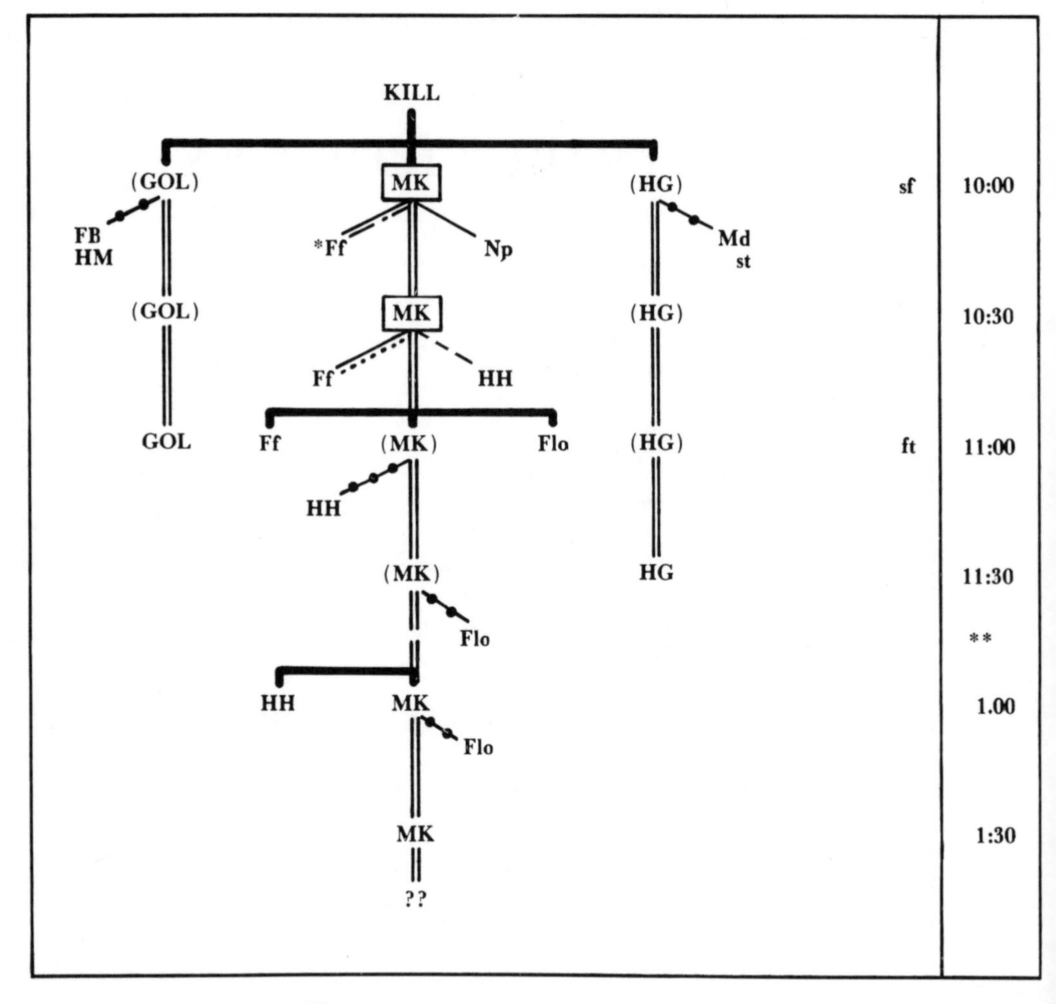

*Estrous females — Ff, waning tumescence
**A 45 minute observation gap.

DIAGRAM II

Stage 3 — April 8, 1968

Meat Distribution

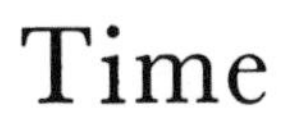

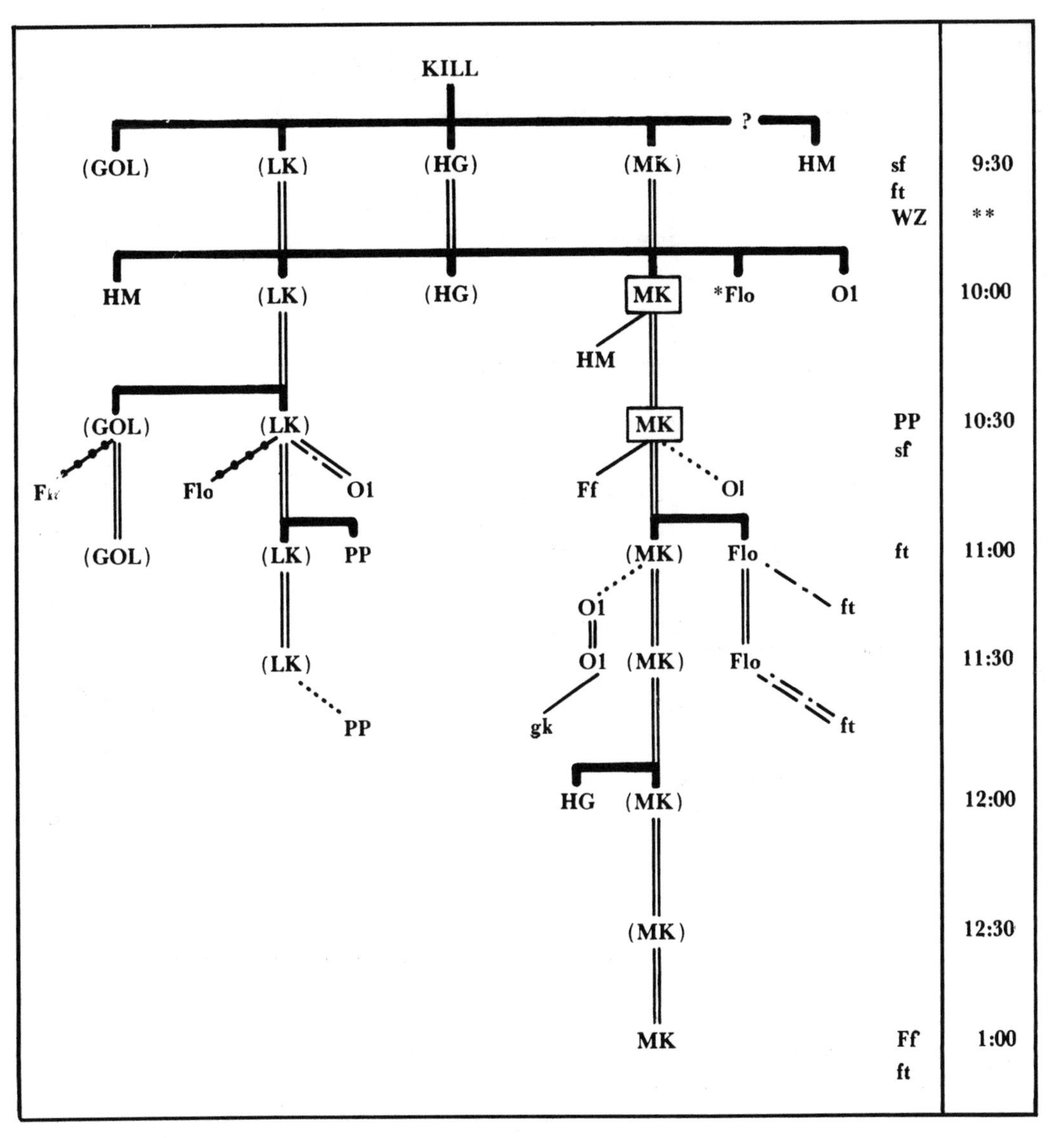

*Estrous females — Flo (pregnant)

**Extensive sharing of meat between MK, GOL, LK, HG, and sometimes Flo and Ol.

DIAGRAM III

Stage 3 — April 15, 1968

Meat Distribution Time

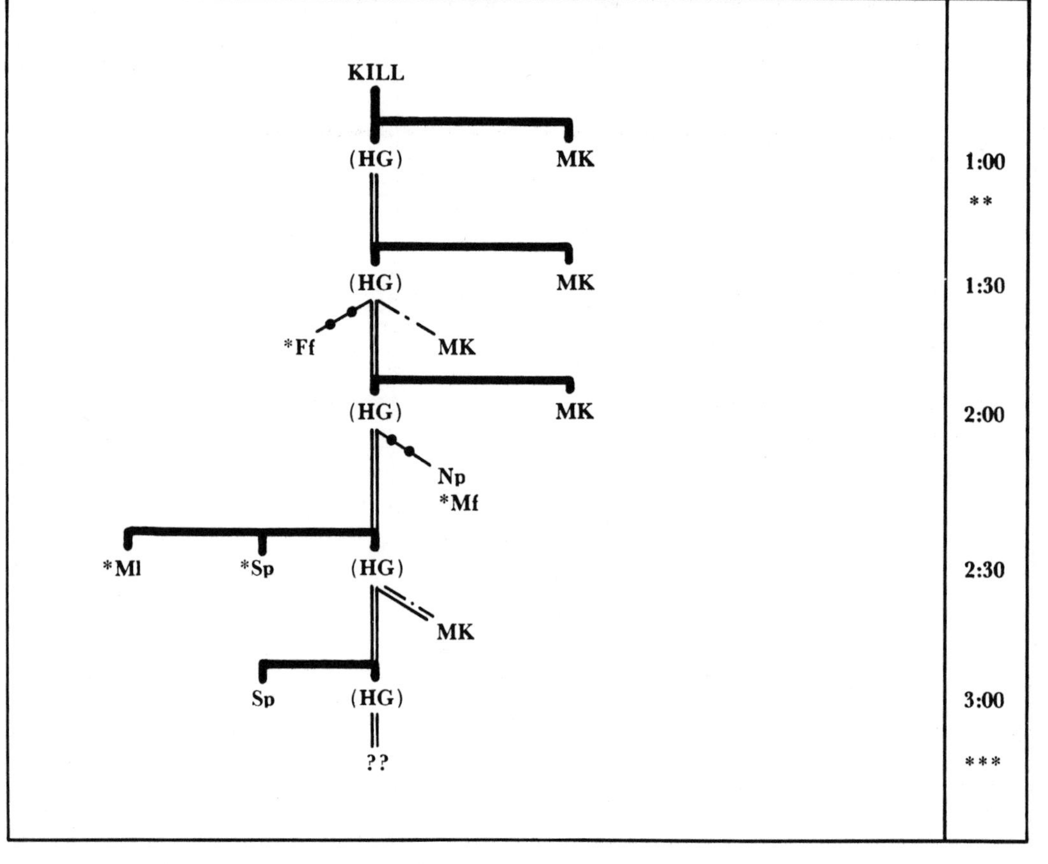

*Estrous females — Ff, waxing tumescence
Mf, Ml, Sp, full tumescence

**Beau baboon stays alive nearly 20 minutes while being eaten.

***Observation discontinued before meat finished.

DIAGRAM IV

Stage 3 — June 19, 1968

Meat Distribution   Time

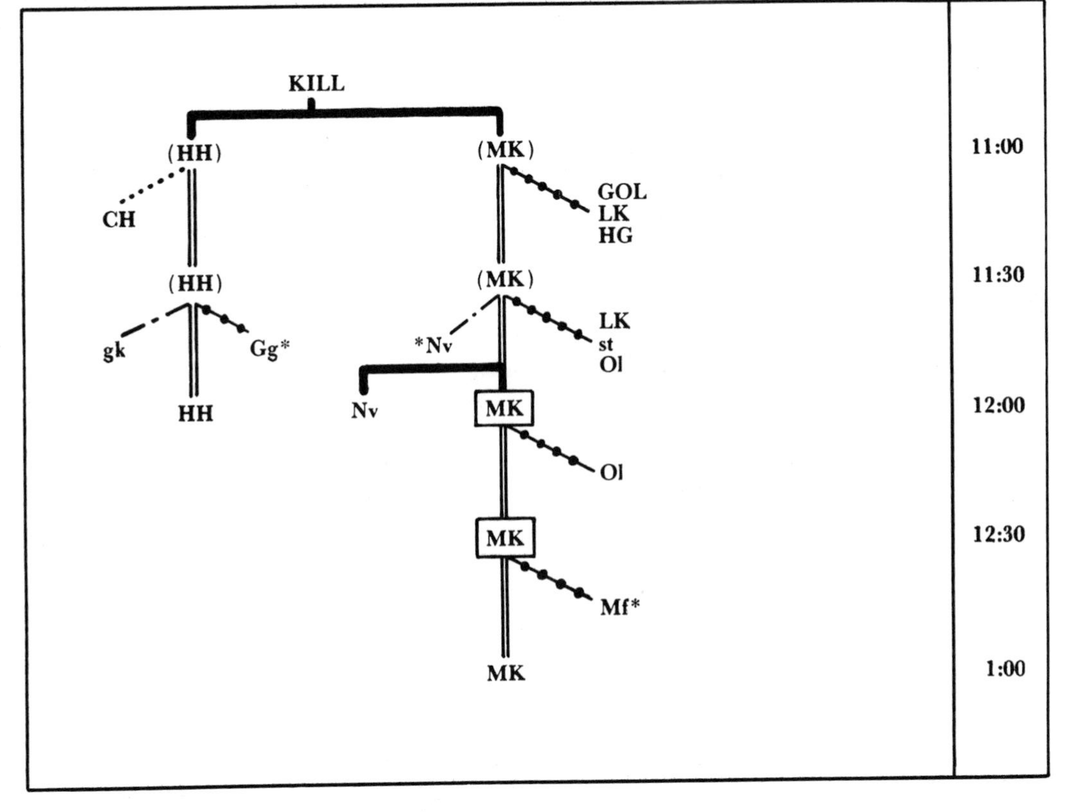

*Estrous females — Gg, Mf, waning tumescence
Nv, waxing tumescence

DIAGRAM V

Stage 3 — June 30, 1968

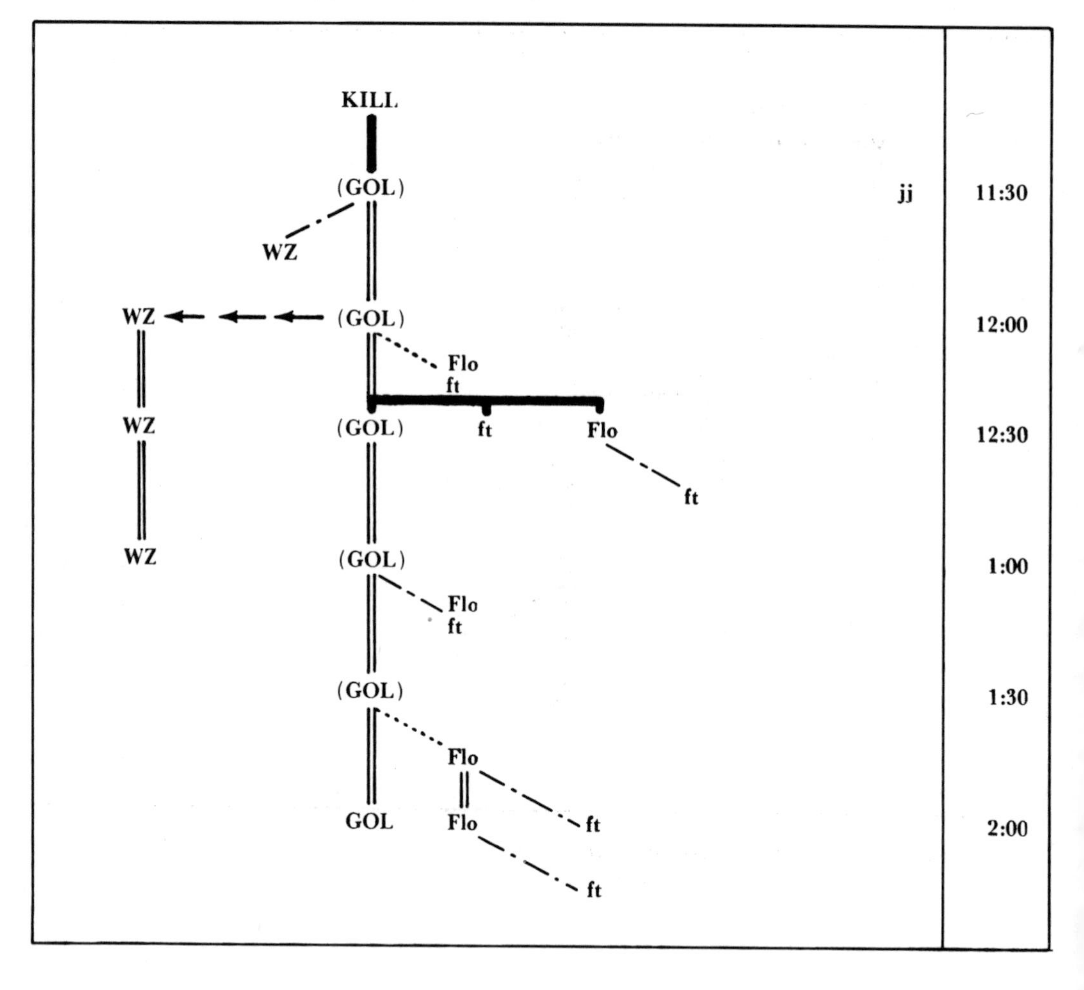

DIAGRAM VI

Stage 3 – Aug. 7, 1968

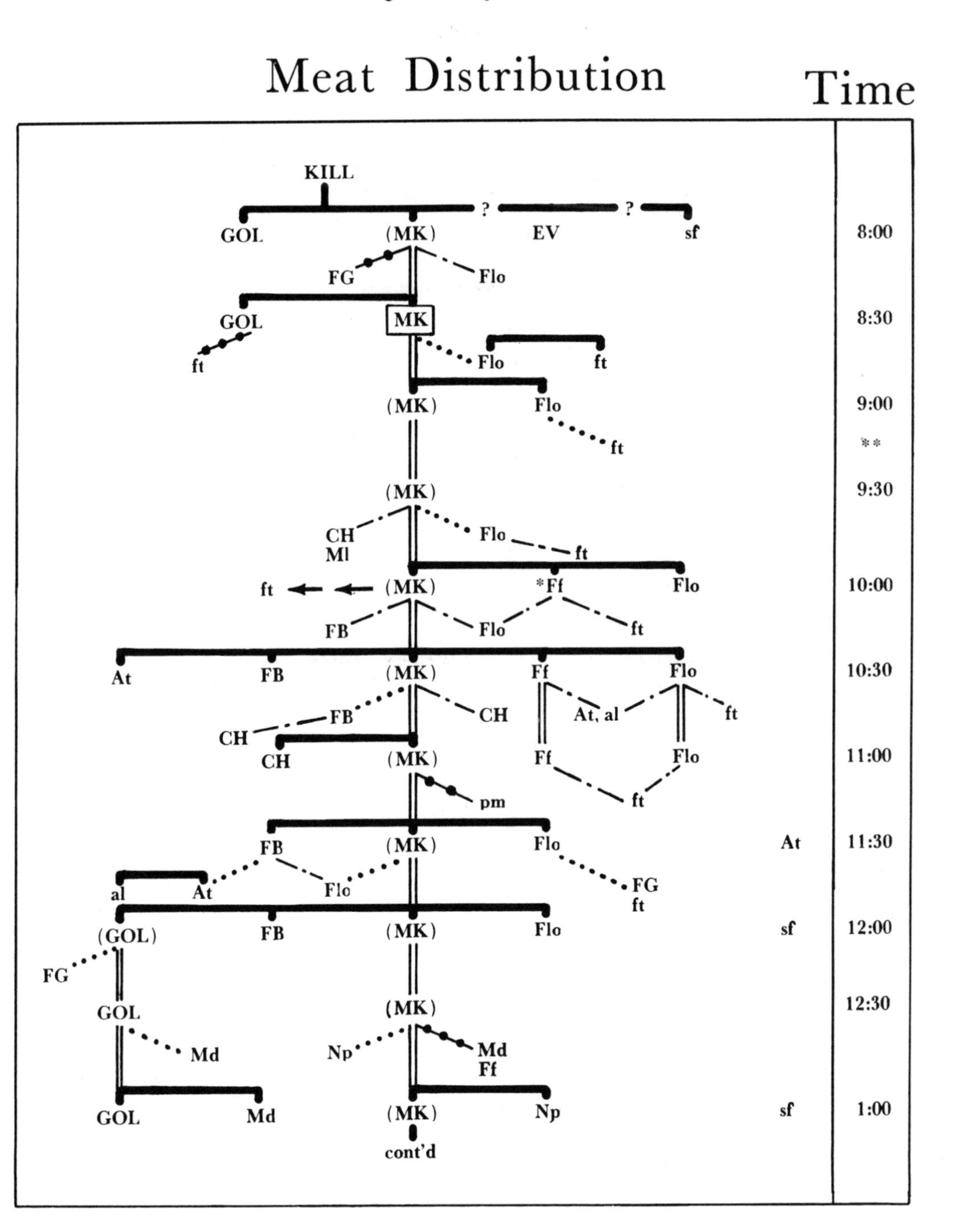

DIAGRAM VI

(cont'd)

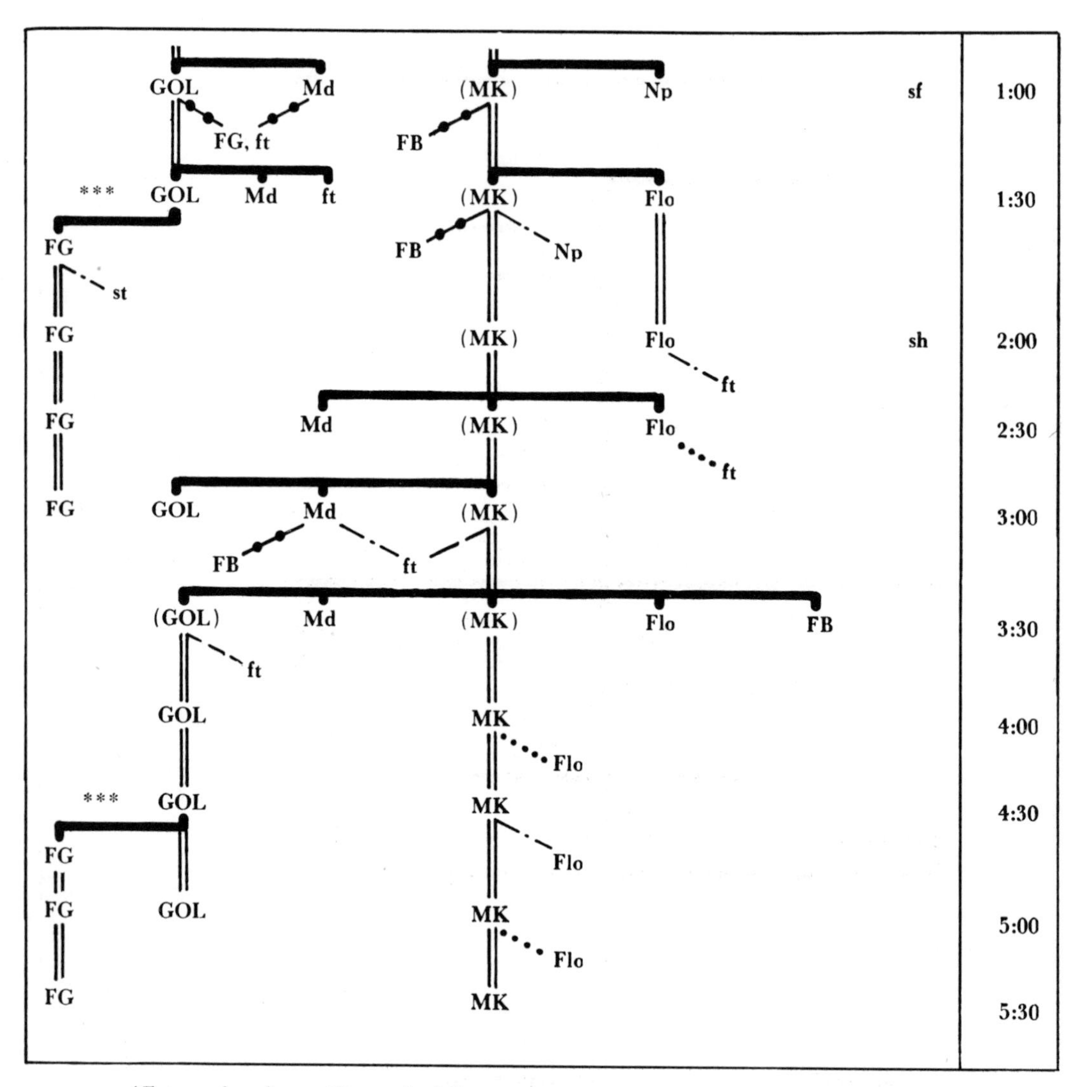

*Estrous females — Ff, nearly full tumescence.

**Chimpanzees return to camp for 30 minutes.

***FG steals meat from GOL with surprise rushes.

DIAGRAM VII

Stage 3 — October 4, 1968

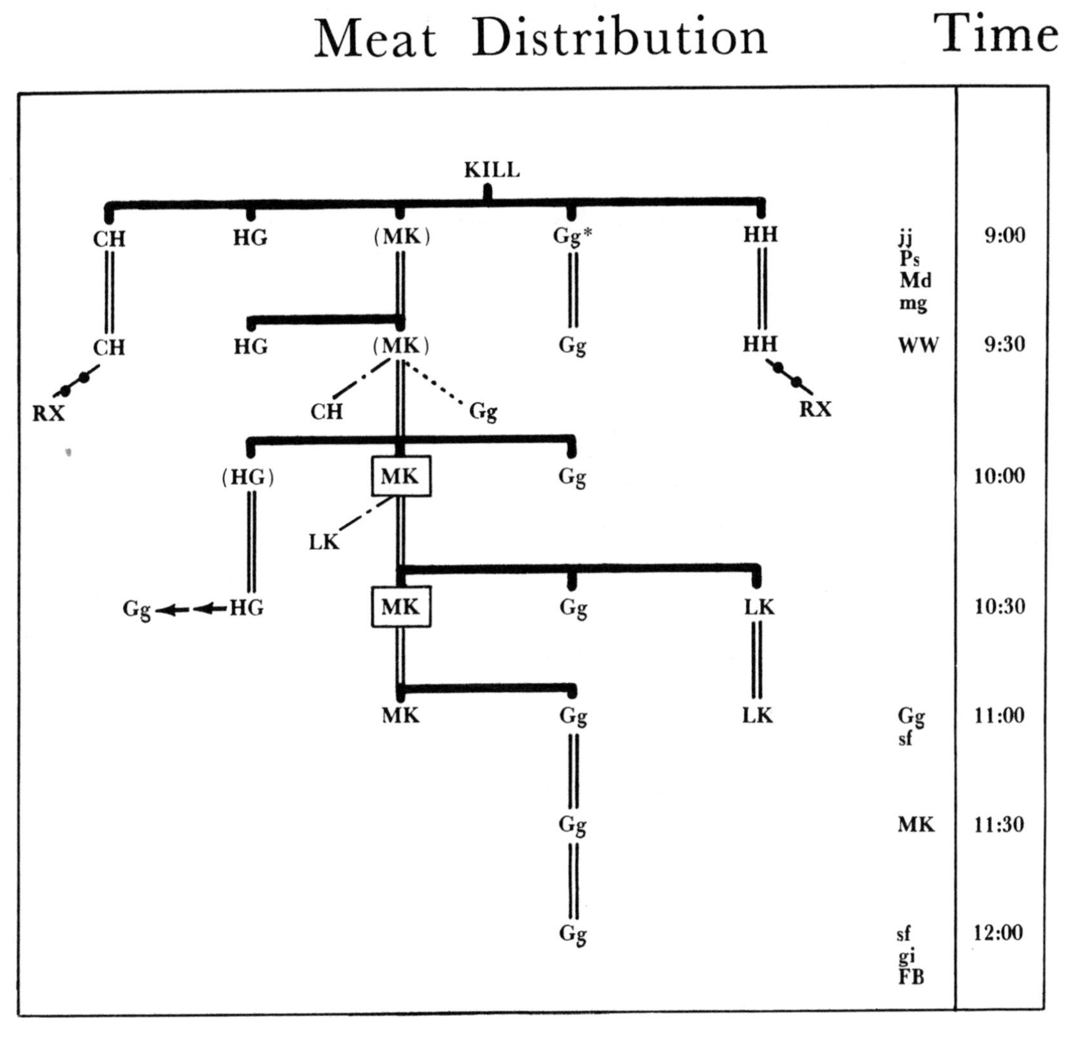

*Estrous females — Gg, nearly full tumescence

DIAGRAM VIII

Stage 3 — December 3, 1968

Meat Distribution Time

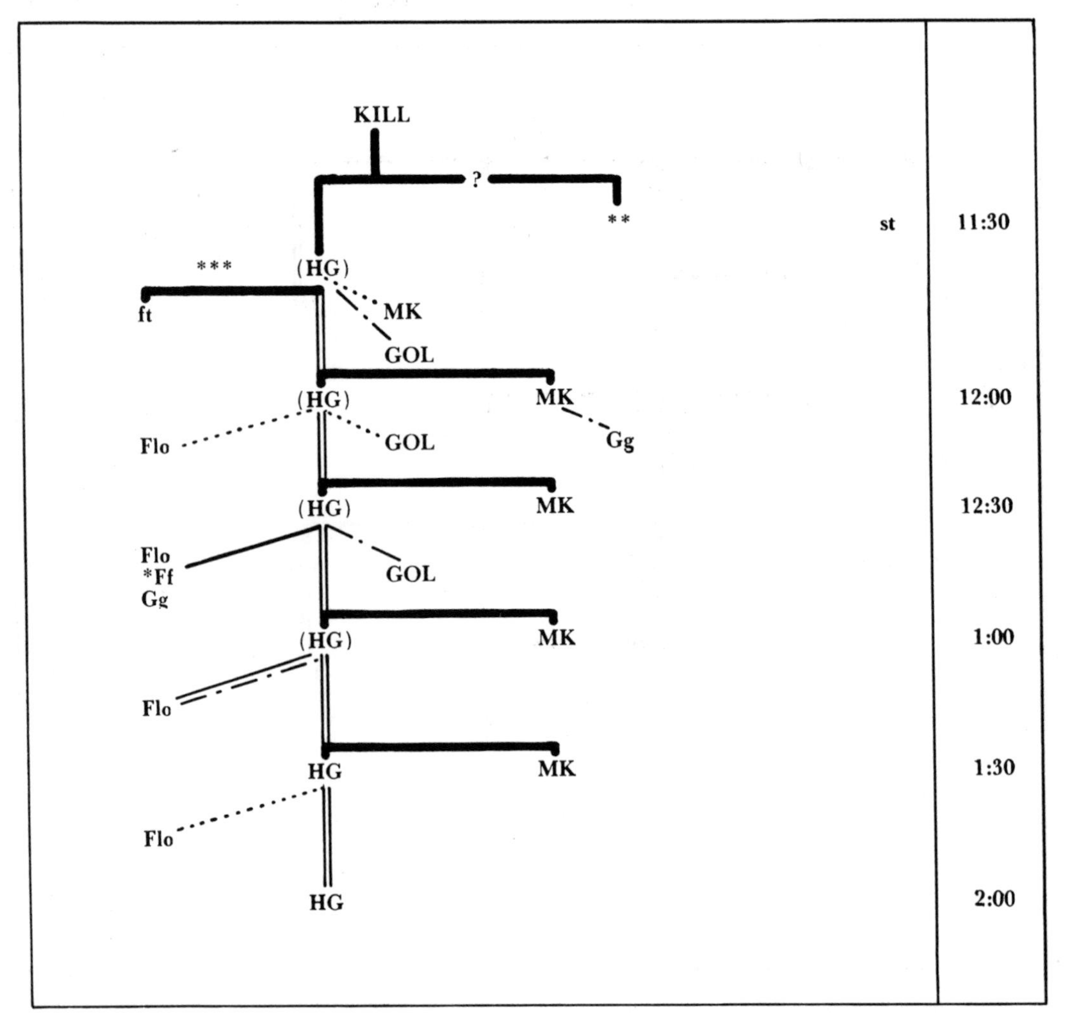

*Estrous females — Ff, waning tumescence
**Baboon head taken, not seen again
***ft sneaks up behind HG and steals a portion

DIAGRAM IX

Stage 3 — Dec. 23, 1968

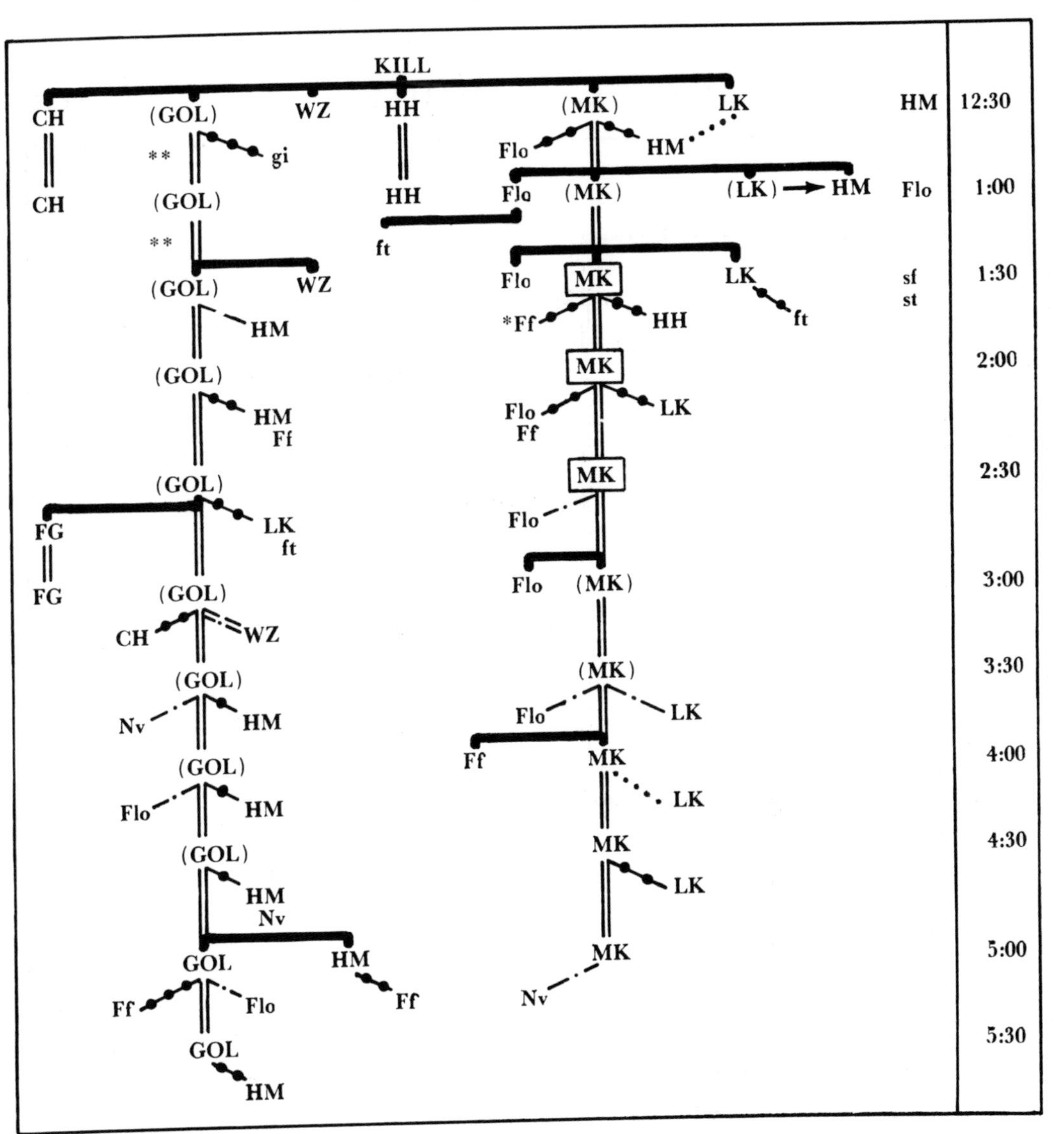

*Estrous females — Ff, nearly full tumescence

**GOL and others who formed a cluster in a nearby tree were not closely observed this hour

DIAGRAM X

Stage 3 — March 11, 1969

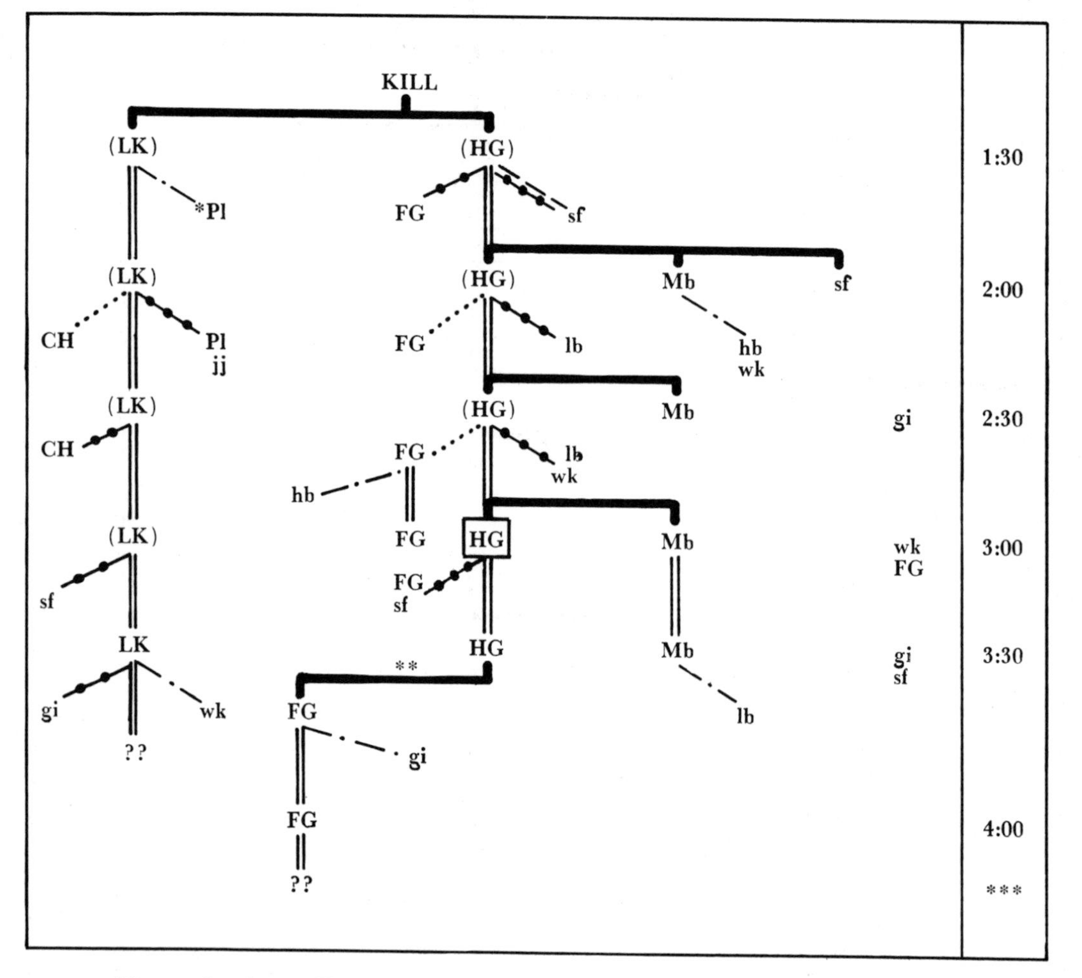

*Estrous females — Pl, full tumescence

**Distracted by general excitement, Hugo drops the carcass; FG retrieves it and moves away

***Chimpanzees disperse

BIBLIOGRAPHY

GENERAL

Altmann, S. A. (ed.), 1967, *Social Communication Among Primates.* Chicago: Univ. Chicago Press.

Ardrey, R., 1961, *African Genesis.* New York: Atheneum.

Ardrey, R., 1966, *The Territorial Imperative.* New York: Atheneum.

Ardrey, R., 1970, *The Social Contract.* New York: Atheneum.

Bartlett, D. and J. Bartlett, 1961, Observations while filming African game. *S. Afr. J. Sci.* 57.

Bates, M., 1958, Food-getting behavior. *Behavior And Evolution* (Roe and Simpson, eds.). New Haven: Yale Univ. Press.

Bernstein, I. S., 1970, Primate status hierarchies. *Primate Behavior: Developments In Field And Laboratory Research* (Rosenblum, ed.). New York: Academic Press.

Carpenter, C. R., 1964, *Naturalistic Behavior Of Nonhuman Primates.* University Park: Pennsylvania State Univ. Press.

Cullen, A., 1969, *Window Onto Wilderness.* Nairobi: East African Pub. House.

Dart, R. A., 1949, The predatory implemental technique of *Australopithecus. Amer. J. Phys. Anthro.* ns 7.

Dart, R. A., 1953, The predatory transition from ape to man. *Intntl. Anthro. Linguist. Rev.* 1.

Dart, R. A., 1955, Cultural status of the South African man-apes. *Smithsonian Ann. Rep.* 317–338.

Dart, R. A., 1957, Osteodontokeratic culture of *Australopithecus prometheus. Transvaal Mus. Mem.* 10.

DeVore, I. (ed.), 1965, *Primate Behavior.* New York: Holt, Rinehart & Winston.

DeVore, I. and S. L. Washburn, 1963, Baboon ecology and human evolution. *African Ecology And Human Evolution* (Howell and Bourliere, eds.). Chicago: Aldine.

Eimerl, S. and I. DeVore, 1965, *The Primates.* New York: Time Inc. (Life Nature Library.)

Estes, R. D. and J. Goddard, 1967, Prey selection and hunting behavior of the African wild dog. *J. Wildlife Management* 31.

Eiseley, L., 1964, *The Unexpected Universe.* New York: Harcourt, Brace & World.

Gibbs, J. L. (ed.), 1965, *Peoples Of Africa.* New York: Holt, Rinehart & Winston.

Harrison, J. L., 1955, Apes and monkeys of Malaya (including the slow loris). *Malayan Mus. Pam.* 9.

Holmes, A., 1965, *Principles Of Physical Geology.* New York: Ronald Press.

Howell, F. C. and F. Bourliere (eds.), 1963, *African Ecology And Human Evolution.* Chicago: Aldine.

Jay, P. (ed.), 1968, *Primates: Studies In Adaptation And Variability.* New York: Holt, Rinehart & Winston.

Kawamura, S. and J. Itani (eds.), 1965, *Monkeys And Apes.* Tokyo: Chuokoran.

Kohler, W., 1925, *The Mentality Of Apes.* New York: Harcourt, Brace & World.

Kruuk, H., 1966, Clan-system and feeding habits of spotted hyenas (*Crocuta crocuta erxleben*). *Nature* 209.

Kruuk, H., 1968, Hyenas. *Nat. Geog.* 134, July.

Kruuk, H. and M. I. M. Turner, 1967, Comparative notes on predation by lion, leopard, cheetah and wild dog in the Serengeti area, East Africa. *Mammalia* 31.

Kuhme, W., 1965, Communal food distribution and division of labor in African hunting dogs. *Nature* 205.

Lawick-Goodall, J. van, 1971, *In The Shadow Of Man.* London: Collins.

Lawick-Goodall, H. and J. van, 1970, *Innocent Killers.* London: Collins.

Leakey, L. S. B., 1969, *Animals Of East Africa.* Washington: National Geographic Society.

Lee, R. B. and I. DeVore (eds.), 1968, *Man The Hunter.* Chicago: Aldine.

LeGros Clark, W. E., 1967, *Man-Apes Or Ape-Men?.* New York: Holt, Rinehart & Winston.

Lorenz, K., 1963, *On Aggression.* New York: Harcourt, Brace & World.

Maberly, C. T. A., 1965, *Animals Of East Africa.* Nairobi: D. A. Hawkins.

Makacha, S., and G. B. Schaller, 1969, Observations on lions in the Lake Manyara National Park, Tanzania. *E. Afr. Wildlife J.* 7.

Marais, E., 1969, *The Soul Of The Ape.* New York: Atheneum.

Marshall, L., 1965, The !Kung bushmen of the Kalahari Desert. *Peoples Of Africa* (Gibbs, ed.). New York: Holt, Rinehart & Winston.

Mech, L. D., 1966, The wolves of Isle Royale. *Fauna of the National Parks of the United States, Fauna Series* 7. Washington: U.S. Government Printing Office.

Montagu, A. (ed.), 1962, *Culture And The Evolution Of Man.* Oxford: Oxford Univ. Press.

Morris, D., 1967a, *The Naked Ape.* New York: McGraw-Hill.

Morris, D. (ed.), 1967b, *Primate Ethology.* Chicago: Aldine.

Morris, D., 1969, *The Human Zoo.* New York: McGraw-Hill.

Morris, R. and D. Morris, 1966, *Men And Apes.* New York: McGraw-Hill.

Mowat, F., 1963, *Never Cry Wolf.* New York: Dell.

Napier, J. R. and P. H. Napier, 1967, *A Handbook Of Living Primates.* New York: Academic Press.

Napier, J. R. and P. H. Napier (eds.), 1970, *Old World Monkeys.* New York: Academic Press.

Oakley, K. P., 1961, On man's use of fire, with comments on tool-making and hunting. *Social Life Of Early Man* (Washburn, ed.). Chicago: Aldine.

Oakley, K. P., 1962, A definition of man. *Culture And The Evolution of Man* (Montagu, ed.). Oxford: Oxford Univ. Press.

Pimlott, D. H., 1967, Wolf predation and ungulate populations. *Amer. Zool.* 7.

Pimlott, D. H., Shannon, J. A. and G. B. Kolensky, 1969, *The Ecology Of The Timber Wolf In Algonquin National Park.* Ontario: Dept. of Lands and Forests.

Reynolds, V., 1965, *Budongo: An African Forest And Its Chimpanzees.* New York: Natural History Press.

Reynolds, V., 1967, *The Apes.* New York: E. P. Dutton.

Roe, A. and G. G. Simpson (eds.), 1958, *Behavior And Evolution.* New Haven: Yale Univ. Press.

Rosenblum, L. A. (ed.), 1970, *Primate Behavior: Developments In Field And Laboratory Research.* New York: Academic Press.

Sahlins, M. D., 1965, The social life of monkeys, apes and primitive man. *The Evolution Of Man's Capacity For Culture* (Spuhler, ed.). Detroit: Wayne State Univ. Press.

Schaller, G. B., 1965a, *The Year Of The Gorilla.* New York: Ballantine.

Schaller, G. B., 1967, *The Deer And The Tiger: A Study Of Wildlife In India.* Chicago: Univ. Chicago Press.

Schaller, G. B., 1968, Hunting behavior of the cheetah in the Serengeti National Park, Tanzania. *E. Afr. Wildlife J.* 6.

Schaller, G. B., 1969a, Life with the king of beasts. *Nat. Geog.* 135, April.

Schaller, G. B., 1969b, Lion ecology and behavior. *Serengeti Res. Inst. Ann. Rep.*

Southwick, C. H. (ed.), 1963, *Primate Social Behavior.* New York: D. Van Nostrand.

Spuhler, J. N. (ed.), 1965, *The Evolution Of Man's Capacity For Culture.* Detroit: Wayne State Univ. Press.

Trewartha, G. T., 1954, *An Introduction To Climate.* New York: McGraw-Hill.

Vagtborg, H. (ed.), 1965, *The Baboon In Medical Research.* Austin: Univ. Texas Press.

Washburn, S. L. (ed.), 1961, *Social Life Of Early Man.* Chicago: Aldine.

Washburn, S. L. and V. Avis, 1958, Evolution of human behavior. *Behavior And Evolution.* (Roe and Simpson, eds.). New Haven: Yale Univ. Press.

Washburn, S. L. and I. DeVore, 1961, Social behavior of baboons and early man. *Social Life Of Early Man* (Washburn, ed.). Chicago: Aldine.

Washburn, S. L. and D. Hamburg, 1968, Aggressive behavior in Old World monkeys and apes. *Primates: Studies In Adaptation And Variability* (Jay, ed.) . New York: Holt, Rinehart & Winston.

Williams, J. G., 1968, *A Field Guide To The National Parks Of East Africa.* Boston: Houghton Mifflin.

Woodburn, J., 1968, An introduction to Hadza ecology. *Man The Hunter* (Lee and DeVore, eds.) . Chicago: Aldine.

Wright, B. S., 1960, Predation on big game in East Africa. *J. Wildlife Management* 24.

Yerkes, R. M., 1943, *Chimpanzees: A Laboratory Colony.* New Haven: Yale Univ. Press.

Yerkes, R. M. and A. W. Yerkes, 1929, *The Great Apes.* New Haven: Yale Univ. Press.

APES

Azuma, S. and A. Toyoshima, 1962, Progress report of the survey of chimpanzees in their natural habitat, Kabogo Point area, Tanganyika. *Primates* 3.

Azuma, S. and A. Toyoshima, 1965, Chimpanzees in Kabogo Point area, Tanganyika. *Monkeys And Apes* (Kawamura and Itani, eds.) . Tokyo: Chuokoran.

Bingham, H. C., 1932, Gorillas in a native habitat. *Carnegie Inst. Wash. Publ.* 462.

Carpenter, C. R., 1938, A survey of wildlife conditions in Atjeh, North Sumatra, with special reference to the orang-utan. *Naturalistic Behavior Of Nonhuman Primates* (Carpenter) . University Park: Pennsylvania State Univ. Press, 1964.

Carpenter, C. R., 1940, A field study in Siam of the behavior and social relations of the gibbon. *Naturalistic Behavior Of Nonhuman Primates* (Carpenter) . University Park: Pennsylvania State Univ. Press, 1964.

Ellefson, J. O., 1967, A natural history of gibbons in the Malay Peninsula. *Ph.D. thesis.* Berkeley: Univ. of California.

Ellefson, J. O., 1968, Territorial behavior in the common white-handed gibbon, *Hylobates lar. Primates: Studies In Adaptation And Variability* (Jay, ed.) . New York: Holt, Rinehart & Winston.

Fossey, D., 1970, Making friends with mountain gorillas. *Nat. Geog.* 137, January.

Fossey, D., 1971, More years with mountain gorillas. *Nat. Geog.* 140, October.

Goodall, J., 1962, Nest building behavior in the free-ranging chimpanzee. *Ann. N.Y. Acad. Sci.* 102.

Goodall, J., 1963a, My life among wild chimpanzees. *Nat. Geog.* 4, August.

Goodall, J., 1963b, Feeding behavior of wild chimpanzees. *Symp. Zool. Soc. London* 10.

Goodall, J., 1964, Tool-using and aimed throwing in a community of free-living chimpanzees. *Nature* 201.

Goodall, J., 1965, Chimpanzees of the Gombe Stream Reserve. *Primate Behavior* (DeVore, ed.) . New York: Holt, Rinehart & Winston.

Harrisson, B., 1962, *Orang-utan*. London: Collins.

Hartmann, R., 1886, *Anthropoid Apes*. New York: D. Appleton.

Itani, J., 1965, Savanna chimpanzees. *Kagahu Asahi* 25.

Itani, J., 1966, The social organization of chimpanzees. *Shizen* 21.

Itani, J., 1967, An artificial feeding of wild chimpanzees and their social organization. *Kagahu Asahi* 27.

Itani, J. and A. Suzuki, 1967, The social unit of chimpanzees. *Primates* 8.

Izawa, K., 1970, Unit groups of chimpanzees and their nomadism in the savanna woodland. *Primates* 11.

Izawa, K. and J. Itani, 1966, Chimpanzees in Kasakati Basin, Tanganyika. *Kyoto: Univ. Afr. Stud.* 1.

Jones, C. and J. Pi Sabater, 1969, Sticks used by chimpanzees in Rio Muni, West Africa. *Nature* 223.

Jones, C. and J. Pi Sabater, 1971, Comparative ecology of *Gorilla gorilla* (Savage and Wyman) and *Pan troglodytes* (Blumenbach) in Rio Muni, West Africa. *Bib. Primat.* 13.

Kawabe, M., 1966, One observed case of hunting behavior among wild chimpanzees living in the savanna woodland of Western Tanganyika. *Primates* 7.

Kawai, M. and H. Mizuhara, 1959, An ecological study of the wild mountain gorilla. (*Gorilla gorilla beringei*). *Primates* 2.

Kortlandt, A., 1962, Chimpanzees in the wild. *Sci. Amer.* 206.

Kortlandt, A., 1963, Bipedal armed fighting in chimpanzees. *XVI Intntl. Congr. Zool. Washington,* lecture text.

Kortlandt, A., 1964, Observation des chimpanzes a l'etat sauvage. *Sci. et Nat.* 61, 65, 66.

Kortlandt, A., 1965, How do chimpanzees use weapons when fighting leopards?. *Yb. Amer. Phil. Soc.* 327–332.

Kortlandt, A., 1966, Chimpansees in vrijheid. *J. Roy. Soc. Zool. Antwerp* 31.

Kortlandt, A., 1967, Experimentation with chimpanzees in the wild. *Progress In Primatology* (Starck, Schneider, and Kuhn, eds.). Stuttgart: Gustav Fischer Verlag.

Kortlandt, A., 1968a, Handgebrauch bei freileben Schimpansen. *Handgebrauch Und Verstandigung Bei Affen Und Fruhmenschen* (Rensch, ed.). Bern: Huber.

Kortlandt, A., 1968b, Die Schlacht der Schimpansen gegen ihren Erbfeind. *Das Tier* 8.

Kortlandt, A., manuscript, Some results of a pilot study on chimpanzee ecology.

Kortlandt, A. and J. C. J. van Zon, 1969, The present state of research on the dehumanization hypothesis of African ape evolution. *Proc. II Intntl. Congr. Primat.* Vol. 3 (Hofer, ed.). New York: S. Karger.

Lawick-Goodall, J. van, 1965, New discoveries among Africa's chimpanzees. *Nat. Geog.* 128, December.

Lawick-Goodall, J. van, 1967a, *My Friends The Wild Chimpanzees.* Washington: National Geographic Society.

Lawick-Goodall, J. van, 1967b, Mother-offspring relationships in free-ranging chimpanzees. *Primate Ethnology* (Morris, ed.). Chicago: Aldine.

Lawick-Goodall, J. van, 1968a, A preliminary report on expressive movements and communication in the Gombe Stream chimpanzees. *Primates: Studies In Adaptation And Variability* (Jay, ed.). New York: Holt, Rinehart & Winston.

Lawick-Goodall, J. van, 1968b, The behaviour of free-living chimpanzees in the Gombe Stream Reserve. *Anim. Beh. Mongr.* 1.

Lawick-Goodall, J. van, 1971, *In the Shadow Of Man.* London: Collins.

Nishida, T., 1967a, Savanna living chimpanzees. *Shizen* 22.

Nishida, T., 1967b, The society of wild chimpanzees. *Shizen* 22.

Nishida, T., 1968, The social group of wild chimpanzees in the Mahali Mountains. *Primates* 9.

Nishida, T., 1970, Social behavior and relationship among wild chimpanzees of the Mahali Mountains. *Primates* 11.

Nissen, H. W., 1931, A field study of the chimpanzee. *Comp. Psychol. Mongr.* 8.

Okano, T., 1965, Preliminary survey of the orang-utan in North Borneo (Sabah). *Primates* 6.

Pfifferling, J. H., in preparation, Demographic trends, 1964–1970, in chimpanzees of the Gombe National Park, Tanzania.

Rensch, B., 1968, *Handgebrauch Und Verstandigung Bei Affen Und Fruhmenschen.* Bern: Huber.

Reynolds, V., 1963, An outline of the behavior and social organization of forest living chimpanzees. *Folia Primat.* 1.

Reynolds, V., 1964, The "man of the woods." *Nat. Hist. N.Y.* 73.

Reynolds, V., 1965a, Ecologie et comportement social des chimpanzees de la foret de Budongo, Ouganda. *Terre et la Vie* 111.

Reynolds, V., 1965b, Some behavioral comparisons between the chimpanzee and the mountain gorilla in the wild. *Amer. Anthro.* 67.

Reynolds, V., 1965c, *Budongo: An African Forest And Its Chimpanzees.* New York: Natural History Press.

Reynolds, V. and F. Reynolds, 1965, Chimpanzees of the Budongo Forest. *Primate Behavior* (DeVore, ed.). New York: Holt, Rinehart & Winston.

Schaller, G. B., 1961, The orang-utan in Sarawak. *Zoologica* 46.

Schaller, G. B., 1963, *The Mountain Gorilla: Ecology And Behavior.* Chicago: Univ. Chicago Press.

Schaller, G. B., 1965a, *The Year Of The Gorilla.* New York: Ballantine.

Schaller, G. B., 1965b, The behavior of the mountain gorilla. *Primate Behavior* (DeVore, ed.). New York: Holt, Rinehart & Winston.

Schaller, G. B. and J. T. Emlen, 1963, Observations on the ecology and social behavior of the mountain gorilla. *African Ecology And Human Evolution* (Howell and Bourliere, eds.). Chicago: Aldine.

Stanley, W. B., 1919, Carnivorous apes in Sierra Leone. *Sierra Leone Stud.* 1.

Sugiyama, Y., 1967, Forest-living chimpanzees. *Shizen* 22.

Sugiyama, Y., 1968, Social organization of chimpanzees in the Budongo Forest, Uganda. *Primates* 9.

Sugiyama, Y., 1969, Social behavior of chimpanzees in the Budongo Forest, Uganda. *Primates* 10.

Suzuki, A., 1966, On the insect-eating habit among wild chimpanzees living in the savanna woodland of Western Tanganyika. *Primates* 7.

Suzuki, A., 1969, An ecological study of chimpanzees in a savanna woodland. *Primates* 10.

Suzuki, A., 1971, Carnivority and cannibalism observed among forest-living chimpanzees. *J. Anthro. Soc. Nippon* 79.

Yerkes, R. M., 1943, *Chimpanzees: A Laboratory Colony.* New Haven: Yale Univ. Press.

Yerkes, R. M. and A. W. Yerkes, 1929, *The Great Apes.* New Haven: Yale Univ. Press.

Yoshiba, K., 1964, Report of the preliminary survey of the orang-utan in North Borneo. *Primates* 5.

Zon, J. C. J. van and J. van Orshoven, 1967, Enkele resultaten van de Zesde Nederlandse Chimpansee-Expeditie. *Vakbl. Biol.* 47.

MONKEYS AND PROSIMIANS

Aldrich-Blake, F., manuscript, The ecology and behavior of the blue monkey *Cercopithecus mitis stuhlmanni. Ph.D. thesis.* Bristol: Univ. of Bristol.

Altmann, S. A., 1959, Field observations on a howling monkey society. *J. Mammal.* 40.

Altmann, S. A., 1962, A field study of the sociobiology of rhesus monkeys. *Ann. N.Y. Acad. Sci.* 102.

Altmann, S. A. (ed.), 1965, *Japanese Monkeys: A Collection Of Translations.* Pub. by the Editor.

Altmann, S. A. and J. Altmann, 1970, *Baboon Ecology: African Field Research.* Basel: S. Karger.

Bernstein, I. S., 1964, A field study of the activities of howler monkeys. *Anim. Beh.* 12.

Bernstein, I. S., 1967, A field study of the pigtail monkey (*Macaca nemestrina*). *Primates* 8.

Bernstein, I. S., 1968, The lutong of Kuala Selangor. *Behav.* 32.

Bolwig, N., 1959, A study of the behavior of the chacma baboon, *Papio ursinus. Behav.* 14.

Bourliere, F., Hunkeler, C. and F. Bertrand, 1970, Ecology and behavior of Lowe's guenon (*Cercopithecus campbelli* Lowei) in the Ivory Coast. *Old World Monkeys And Apes* (Napier and Napier, eds.). New York: Academic Press.

Buxton, A. P., 1952, Observations on the diurnal behavior of the redtail monkey (*Cercopithecus ascanius schmidti* Matsche) in a small forest in Uganda. *J. Anim. Ecol.* 21.

Carpenter, C. R., 1934, A field study of the behavior and social relations of howling monkeys. *Naturalistic Behavior Of Nonhuman Primates* (Carpenter). University Park: Penn. State Univ. Press, 1964.

Carpenter, C. R., 1935, Behavior of the red spider monkey (*Ateles geoffroyi*) in Panama. *Naturalistic Behavior Of Nonhuman Primates* (Carpenter). University Park: Pennsylvania State Univ. Press, 1964.

Carpenter, C. R., 1965, The howlers of Barro Colorado Island. *Primate Behavior* (DeVore, ed.). New York: Holt, Rinehart & Winston.

Chalmers, N. R., 1967, The ecology and ethology of the black mangabey *Cercocebus albigena. Ph.D. thesis.* Cambridge: Univ. of Cambridge.

Chalmers, N. R., 1968a, Group composition, ecology and daily activities of free-living mangabeys in Uganda. *Folia Primat.* 8.

Chalmers, N. R., 1968b, The social behavior of free-living mangabeys in Uganda. *Folia Primat.* 8.

Clutton-Brock, T., in preparation, (The feeding ecology of red colobus monkeys in Gombe National Park.)

Dart, R. A., 1963, The carnivorous propensity of baboons. *Symp. Zool. Soc. London* 10.

DeVore, I. and K. R. L. Hall, 1965, Baboon ecology. *Primate Behavior* (DeVore, ed.). New York: Holt, Rinehart & Winston.

DeVore, I. and S. L. Washburn, 1963, Baboon ecology and human evolution. *African Ecology And Human Evolution* (Howell and Bourliere, eds.). Chicago: Aldine.

Durham, N. M., in preparation, A preliminary description of the naturalistic behavior of brown woolly monkeys, *Lagothrix lagothrica.*

Furuya, Y., 1961–62, The social life of silvered leaf monkeys (*Trachypithecus cristatus*). *Primates* 3.

Furuya, Y., 1962, Ecological survey of the wild crab-eating monkeys of Malaya. *Primates* 3.

Gartlan, J. S., 1968, Ecology and behavior of an isolated population of vervet monkeys on Lolui Island, Lake Victoria. *Human Biol.* 40.

Gartlan, J. S., 1970, Preliminary notes on the ecology and behavior of the drill. *Mandrillus leucophaeus* Ritgen 1824. *Old World Monkeys* (Napier and Napier, eds.). New York, Academic Press.

Gartlan, J. S. and C. K. Brain, 1968, Ecology and social variability in *Cercopithecus aethiops* and *C. mitis. Primates: Studies In Adaptation And Variability* (Jay, ed.). New York: Holt, Rinehart & Winston.

Haddow, A. J., 1956, The blue monkey group in Uganda. *Uganda Wildlife and Sport* 1.

Hall, K. R. L., 1961, Feeding habits of the chacma baboon. *Adv. Sci.* 17.

Hall, K. R. L., 1962a, Numerical data, maintenance activities, and locomotion of the wild chacma baboon. *Proc. Zool. Soc. London* 139.

Hall, K. R. L., 1962b, The sexual, agonistic and derived social behavior patterns of the wild chacma baboon, *Papio ursinus. Proc. Zool. Soc. London* 139.

Hall, K. R. L., 1963, Variation in the ecology of the chacma baboon, *Papio ursinus. Symp. Zool. Soc. London* 10.

Hall, K. R. L., 1965, Ecology and behavior of baboons, patas and vervet monkeys. *The Baboon In Medical Research* (Vagtborg, ed.). Austin: Univ. Texas Press.

Hall, K. R. L., 1966, Distribution and adaptations of baboons. *Symp. Zool. Soc. London* 17.

Hall, K. R. L. and I. DeVore, 1965, Baboon social behavior. *Primate Behavior* (DeVore, ed.). New York: Holt, Rinehart & Winston.

Hall, K. R. L. and J. S. Gartlan, 1965, Ecology and behavior of the vervet monkey, *Cercopithecus aethiops,* Lolui Island, Lake Victoria. *Proc. Zool. Soc. London* 145.

Imanishi, K., 1957, Social behavior in Japanese monkeys, *Macaca fuscata. Psychologia* 1.

Jay, P., 1963, The Indian langur monkey *(Presbytis entellus)*. *Primate Social Behavior* (Southwick, ed.). New York: D. Van Nostrand.

Jay, P., 1965, The common langur of North India. *Primate Behavior* (DeVore, ed.). New York: Holt, Rinehart & Winston.

Jolly, A., 1966, *Lemur Behavior: A Madagascar Field Study.* Chicago: Univ. Chicago Press.

Kern, J. A., 1964, Observations on the habits of the proboscis monkey, *Nasalis larvatus* (Wurmb), made in the Brunei Bay area, Borneo. *Zoologica* 49.

Kern, J. A., 1965, The proboscis monkey. *Anim. Kingd.* 68.

Koford, C. B., 1963, Group relations in an island colony of rhesus monkeys. *Primate Social Behavior* (Southwick, ed.). New York: D. Van Nostrand.

Kummer, H., 1968, *Social Organization Of Hamadryas Baboons: A Field Study*. Chicago: Univ. Chicago Press.

Kummer, H. and F. Kurt, 1963, Social units of a free-living population of hamadryas baboons. *Folia Primat.* 1.

Marais, E., 1969, *The Soul Of The Ape*. New York: Atheneum.

Maxim, P. E. and J. Buettner-Janusch, 1963, A field study of the Kenya baboon. *Amer. J. Phys. Anthro.* 21.

Neville, M. K., 1968, Ecology and activity of Himalayan foothill rhesus monkeys. *Ecology* 49.

Oppenheimer, J. R., 1968, Behavior and ecology of the white-faced monkey, *Cebus capucinus,* on Barro Colorado Island. *Ph.D. thesis.* Univ. of Illinois.

Petter, J. J., 1962, Ecological and behavioral studies of Madagascar lemurs in the field. *Ann. N.Y. Acad. Sci.* 102.

Petter, J. J., 1965, The lemurs of Madagascar. *Primate Behavior* (DeVore, ed.). New York: Holt, Rinehart & Winston.

Poirier, F. E., 1970, The Nilgiri langur *(Presbytis johnii)* of South India. *Primate Behavior: Developments In Field And Laboratory Research* (Rosenblum, ed.). New York: Academic Press.

Ransom, T. W., 1971, Ecology and social behavior of baboons (*Papio anubis*) at the Gombe National Park. *Ph.D. thesis.* Berkeley: Univ. California.

Ransom, T. W. and B. S. Ransom, in press, Special relationships among adult male and immature baboons and the formation of social bonds.

Ripley, S., 1965, The ecology and social behavior of the Ceylon gray langur (*Presbytis entellus*). *Ph.D. thesis.* Berkeley: Univ. of California.

Rowell, T. E., 1966, Forest-living baboons in Uganda. *J. Zool. London* 149.

Simonds, P. E., 1965, The bonnet macaque in South India. *Primate Behavior* (DeVore, ed.). New York: Holt, Rinehart & Winston.

Struhsaker, T., 1967a, Ecology of vervet monkeys (*Cercopithecus aethiops*) in the Masai-Amboseli Game Reserve, Kenya. *Ecology* 48.

Struhsaker, T., 1967b, Auditory communication among vervet monkeys (*Cercopithecus aethiops*). *Social Communication Among Primates* (Altmann, ed.). Chicago: Univ. Chicago Press.

Sugiyama, Y., 1967, Social organization of hanuman langurs. *Social Communication Among Primates* (Altmann, ed.). Chicago: Univ. Chicago Press.

Washburn, S. L. and I. DeVore, 1961a, Social behavior of baboons and early man. *Social Life Of Early Man* (Washburn, ed.). Chicago: Aldine.

Washburn, S. L. and I. DeVore, 1961b, The social life of baboons. *Sci. Amer.* June (Reprint).

Yoshiba, K., 1967, An ecological study of hanuman langurs (*Presbytis entellus*). *Primates* 8.

Yoshiba, K., 1968, Local and intertroop variability in ecology and social behavior of common Indian langurs. *Primates: Studies In Adaptation And Variability* (Jay, ed.). New York: Holt, Rinehart & Winston.

MAPS

East Africa, Shell Road Map of—1965, scale 1:2,000,000; London: George Phillip & Son.

Kigoma, Map of—1961, scale 1:125,000, Quarter Degree Sheet 92; Dodoma: Geological Survey Division.

Tabora, Map of—1964, scale 1:1,000,000, International Map of the World Series 1301, edition 3-DOS; Director of Overseas Surveys for Republic of Tanganyika and Zanzibar.

Tanzania, Map of—scale 1:2,000,000, 4th edition; Survey Division, Ministry of Lands, Settlement and Water.

FILMS*

Man:	The Hunters (70003N)
Gorilla:	Mountain Gorilla (PCR 2141K)
Chimpanzee:	Miss Goodall and the Wild Chimpanzees (31269)
Baboon:	Animals in Amboseli (20773)
	Baboon Behavior (PCR 2107K)
	Chacma Baboons: Ecology and Behavior (PCR 2167)
	Dynamics of Male Dominance in a Baboon Troop (31292)
	The Young Infant: Birth to Four Months (10355)

* These and other films on primate behavior are listed in "Films: The Visualization of Anthropology," a catalog compiled by L. Carpenter and J. H. Pfifferling. Numbered films are available from: Audio-Visual Services, 6 Willard Building, Pennsylvania State University, University Park, Pa. 16802.

The Older Infant: Four Months to One Year (10353)

Macaque: Behavior of the Macaques of Japan: The *Macaca fuscata* of the Takasakiyama and Koshima Colonies (PCR 2184K)

Behavioral Characteristics of the Rhesus Monkey (PCR 2011)

Rhesus Monkeys in India (PCR 2129K)

Social Behavior of Rhesus Monkeys (PCR 2012)

INDEX

HUMAN NAMES

CHIMPANZEE NAMES

SUBJECTS